The Art of
Psychological Warfare

1914–1945

1. Anti-German atrocity propaganda, linked personally with the Kaiser.

The Art of
Psychological Warfare
1914–1945

CHARLES ROETTER

STEIN AND DAY / *Publishers* / New York

First published in the United States of America in 1974
Copyright © 1974 by Charles Roetter
Library of Congress Catalog Card No. 74-80901
All rights reserved
Printed in the United States of America
Stein and Day/*Publishers*/Scarborough House,
    Briarcliff Manor, N.Y. 10510
ISBN 0-8128-1737-0

Contents

List of Illustrations

Acknowledgments

The Author and Publisher would like to thank the Imperial War
Museum for permission to use the illustrations appearing in this book.

Preface

Propaganda is to many a suspect word and a suspect activity. It smacks of trying to make people do things against their will, of under-hand methods, of sharp practice, of deceit, of trickery.

No doubt propaganda has been and is guilty of all these charges. After all, a form of activity which seeks to influence people's minds and attitudes and, if possible, action is bound at times to lead its practitioners astray. Whether their work gains in effectiveness as a result, is another question. For when truth itself becomes a commodity which is in short supply or non-existent, it can become a more powerful method of propaganda than any other. In times of confusion many people would rather know what is happening than be fobbed off with cleverly contrived explanations or interpretations.

Propaganda and psychological warfare are as old as the history of mankind itself. The weak have always sought to make themselves out stronger than they were. Sometimes they were successful and sometimes they were not. The Byzantine Empire, in its efforts to prolong its existence, resorted to such devices as having the emperor on his throne raised as if by divine intervention and lowered again while the stuffed lions which served as the arm-rests of his throne belched forth fire and smoke and uttered terrifying roars as the ambassadors from foreign lands lay prostrate before him. It also, from time to time, organized for the diplomatic corps, military parades in which the Imperial Guard – as soon as it had disappeared from sight of the reviewing stand – hastily changed uniforms and re-appeared again, seeking to give the illusion of being but a fraction of an army numbering hundreds of thousands.

Since all foreign diplomats, indeed all foreigners, were kept under strict surveillance in ancient Byzantium such ruses may have been effective for a time. They could not prevent the ultimate collapse and disappearance of the Byzantine Empire.

Armour and uniforms have always been designed to give pride, courage and confidence to its wearer and to strike terror in the heart of the enemy. When primitive warriors put on war paint, it was not only to protect themselves by magic symbols but to intimidate the enemy by showing him plainly that they meant business – death and destruction.

One of the most effective devices of dual-purpose propaganda was the use of drums by Red Indian tribes before attacking a white settlement or wagon train. The drums aroused the fighting spirit of the Red Indian warriors while their sound gnawed away at the nerves of the white settlers who knew not when the drums would stop, but knew that once they did, the attack was due to begin soon.

The Christian Churches, indeed all proselytizing religions, have never regarded propaganda with disfavour. On the contrary, they have always felt it an obligation to propagate their faith with all the power at their command. The great political ideologies take a similar view.

In any case, whether or not we look upon some manifestations of propaganda with distaste, propaganda has become an essential element in the relations between states and within states. Wars are no longer the business of professional armies nor politics the exclusive preserve of relatively small oligarchies. This century has seen two world wars. They have in every sense been "total", drawing in and absorbing the energies of every section of the warring nations. Their attitudes, their moods, played a vital part on the way in which they acted and conducted themselves. And in shaping their attitudes, those moods, their own propaganda and that of their enemies exerted a key influence.

This book will examine the principles, the ground rules by which practioners of propaganda ought to be guided, and it will look at how these principles, these ground rules have been applied in two world wars, and with what success.

Charles Roetter

The Art of
Psychological Warfare
1914–1945

CHAPTER 1 # Ground Rules for Propaganda

In August 1915 Nurse Edith Cavell, who had been working at a Red Cross hospital in German-occupied Brussels, was found guilty by a German court martial of helping British and French soldiers to escape into neutral Holland. She was sentenced to death and executed by a German firing squad.

British public opinion was outraged. More important from the polical point of view was the wave of revulsion that swept across the United States when the news of her execution became known. Certainly the British authorities concerned with trying to win over a United States which was then still neutral, to the Allied side, spared no effort to ensure that Nurse Edith Cavell's execution was given the widest publicity across the length and breadth of America. Even American newspapers which were at that stage of the First World War inclined to be pro-German or at least sceptical of the justice of the Allied cause, printed full accounts of the 50-year-old devoted nurse's calm and courageous composure as she faced the German firing squad. Many Americans who had been suspicious of the atrocity stories about German behaviour in Belgium and had been inclined to dismiss them as 'clever propaganda' by the British, were suddenly saying that the British might even have been guilty of their traditional tendency of understatement. Plainly the German occupation of Belgium was worse and more cruel than anyone had imagined. To execute a woman – and to execute her by a military firing squad – that was not the behaviour of a civilized nation. So perhaps all the other stories about children being roasted on soldiers' bayonets over camp fires were true as well after all?

Another 20 months were to pass before the United States eventually decided to enter the First World War on the side of Britain and her Allies against Imperial Germany and the Central Powers, but in retrospect Nurse Cavell's execution provoked one of the most significant reactions among ordinary Americans in favour of the Allies and against Germany.

By itself it might soon have been forgotten, but the emotional impact it caused was sustained and kept alive by the imposition by Germany of unlimited submarine warfare in order to break Britain's naval blockade of Germany, a campaign Germany had to abandon when the *Lusitania* was in 1915 sunk by a German submarine with the loss of many American lives. Unfortunately for Imperial Germany, the German High Sea Fleet failed to destroy the British Navy at Jutland in 1916. The battle was a 'stand-off', and the German High Sea Fleet never left port again. The result was that the British naval blockade of Germany continued and became steadily more effective.

By the beginning of 1917 the Imperial German High Command felt that it had no choice but to resume unrestricted U-boat warfare regardless of the consequences. In April of the same year, the United States entered the War on the side of the Allies.

In short, the cumulative effect of the way Imperial Germany behaved was to project an image to many Americans of a country prepared to flout any moral and civilized mode of conduct. It executed nurses by military firing squad; it sunk, without prior warning, civilian ships like the *Lusitania* carrying dozens of American citizens whose country was neutral; and although it temporarily stopped these wicked and illegal attacks after long and sustained protests from the neutral United States, it resumed them – although they were deemed wrong in international law – the moment it felt them to be in its own self-interest. Here surely was a tragic example of inhuman and barbaric conduct. America had no choice but to come down on the side of righteousness!

In retrospect it is not difficult to assess the crucial part Nurse Cavell's execution played in influencing public opinion towards the two belligerent camps in the First World War, particularly in neutral countries like the United States. But even without the benefit of hindsight, were the Germans themselves at the time entirely blind to the serious consequences of actually carrying out the sentence of the court martial?

There is evidence that some Germans at least were not. The only man who could commute a sentence of death passed by a court martial was the Kaiser in his capacity as 'Supreme War Lord'. A request to that effect was passed on to him, and it had the support of no less a personage than the German military governor of occupied Belgium. It would be wrong to suggest that the military governor, in supporting this request, was motivated by the possible effect the execution might have on American public opinion. He was in all likelihood probably much more concerned with its effect on Belgian public opinion. It would be equally wrong to exclude the possibility that the military governor may at least to some extent have been prompted by common human decency.

One aspect of the whole affair is that there is even now no conclusive evidence of what happened to the request for mercy. Did it ever reach the Kaiser, or was it mysteriously held up somewhere on its passage up to the hierarchical top of the German war machine? And if it reached the Kaiser, what was his reaction to the advice tendered him by his entourage? We do not know. All we do know, is that Nurse Cavell was executed.

Some weeks after her execution two German nurses working in a Red Cross hospital in France were found guilty by a French court martial of helping German prisoners of war escape and sentenced to death. Both of them were subsequently executed by a French firing squad. Their execution was given only a brief mention in a few French, German and European neutral newspapers. It appears not to have been reported at all in the American press.

When an American correspondent, based in Berlin, asked one of the German officers at the War Ministry responsible for propaganda to the United States, why after the sense of outrage which Nurse Cavell's execution had aroused in America, the German propagandists were making no effort to retaliate, not even by at least bringing the event to the attention of the American public, he met with total incomprehension. The German nurses, the officer explained, had acted in contravention of the rules of the Red Cross; the French court martial – like the German court martial in Nurse Cavell's case – had acted according to the rules of war; the two German nurses – like Nurse Cavell – had behaved bravely and patriotically and they had paid the penalty for acting outside the rules of war. Surely, he concluded, there was no case for trying to score propaganda points out of a situation like that.

Plainly the German officer in this case was entirely unsuited for his work, and it is doubtful if even months of training would have made him understand what propaganda and psychological warfare are about. Certainly he completely ignored one of the first and most important ground rules in propaganda: to be absolutely clear about his objective.

No one who wishes to make propaganda effectively and conduct psychological warfare, no one, in other words, who wishes to influence other people's attitudes and possibly actions, can hope to succeed unless he is absolutely clear of what he wants those people to think and to do. On this there can be no vagueness, no woolliness. His objective, his aim, what it is he wishes to achieve, must be as precise as possible. His aim may be to boost home morale. It may be to sow distrust between the enemy's generals and ordinary soldiers. It may be to demoralize the enemy's fighting men by making them feel that their families at home are not receiving adequate rations, or that those at the top of the social scale are getting better treatment than those at the bottom. Or that their equipment is not as good as that of their opponents, that it is good sense to surrender now and survive with a chance of rejoining their families rather than fight on and get caught up in the general holocaust.

The objective, the aim in the case of the German officer concerned with propaganda to the United States, was simply and plainly to keep America neutral, to stop her from entering the war on the side of Britain and her Allies. Nurse Cavell's execution by a German firing squad may by itself not have sent thousands of Americans rushing across the Atlantic to fight the Kaiser's armies, but it hardly served to enhance Germany's claim to be a civilized nation with a long tradition of *Kultur*. The execution of the two German nurses by the French could have been used at least to point up the obvious truth that war can be and frequently is rough and cruel. The German officer's comment that all three executions were in accordance with the rules of war would, if it had been given wide publicity at the time, have served only to convince many Americans of the lack of humanity of the Kaiser's Germany.

This comment also points up the officer's ignorance of another of the important ground rules in propaganda: a clear idea and knowledge of the target which is being aimed at. In other words, he had failed to do a 'target analysis' of the United States. It was entirely irrelevant that in his personal view and in the light of his personal social background he may – as he obviously did – feel that many Americans were being sentimental about having women executed by firing squad. The point was that most Americans – male and female – did not like the idea, and that Nurse Edith Cavell's execution helped to build up the image of the Kaiser's Germany as a barbaric country. It was that image that he should have been trying to fight. If he had known his business as a propagandist, if he had been at all efficient as a psychological warrior – which plainly was not the case – he would have known that no matter how strongly he might have felt on the transgressions of the rules of warfare and the limits on the immunity afforded to nurses by the Red Cross, he had no chance of putting

2. 'Brave little Belgium' was the theme of dozens of posters designed to arouse Britain's fighting spirit and generosity.

3. The sinking of the *Lusitania* with the loss of many American lives provided the theme for appeals to then neutral America to join the War against Germany.

DON'T IMAGINE YOU ARE NOT WANTED

EVERY MAN between 19 and 38 years of age is WANTED! Ex-Soldiers up to 45 years of age

MEN CAN ENLIST IN THE NEW ARMY FOR THE DURATION OF THE WAR

"YOUR COUNTRY NEEDS YOU"

RATE OF PAY: Lowest Scale 7s. per week with Food, Clothing &c., in addition

1. Separation Allowance for Wives and Children of Married Men when separated from their Families (Inclusive of the allotment required from the Soldier's pay of a maximum of 6d. a day in the case of a private)

For a Wife **without** Children -	12s. 6d. per week
For Wife with One Child -	15s. 0d. per week
For Wife with Two Children -	17s. 6d. per week
For Wife with Three Children -	20s. 0d. per week
For Wife with Four Children -	22s. 0d. per week

and so on, with an addition of 2s. for each additional child.
Motherless children 3s. a week each, exclusive of allotment from Soldier's pay

2. Separation Allowance for Dependants of Unmarried Men.

Provided the Soldier does his share, the Government will assist liberally in keeping up, within the limits of Separation Allowance for Families, any regular contribution made before enlistment by unmarried Soldiers or Widowers to other dependants such as mothers, fathers, sisters, etc.

YOUR COUNTRY IS STILL CALLING.
FIGHTING MEN!   FALL IN!!

Full Particulars can be obtained at any Recruiting Office or Post Office.

No 0200
DAVID ALLEN & SONS L^TD
HARROW
LONDON

4. A British recruiting poster with Alfred Leete's famous 'Kitchener' as its centrepiece.

his case effectively to the American public in the time available to him.

This was a mistake the Allied propagandists in the Second World War did not make in trying to secure some kind of support or connivance from German officers. Allied propagandists knew that all German officers had been made to swear a personal oath of allegiance to Hitler. Moreover, they knew that an oath of that kind was deeply bound up with personal honour and integrity as a professional soldier. Therefore to argue the ethics of the circumstances in which an officer and a gentleman could consider himself released from such an oath would in many cases take more time than finishing the war. Other arguments had therefore to be used as well – and they were.

Target analysis – a term better known in the Second World War than in the First – was, however, well understood by a great many Germans in the First Great War. One of those Germans was Count Bernstorff, Imperial Germany's ambassador in Washington while the U.S. remained neutral in the 1914–18 War. Unlike the German officer in the War Ministry in Berlin, he had made his 'target analysis'. He knew that Americans did not take to nations who formally, with the full panoply of the law, had their soldiers execute women or used their armed forces to starve children.

Britain's naval blockade of Germany and the Central Powers was by 1916 beginning to bite. In Germany there were meatless days, fresh vegetables had to be rationed and so had milk and butter, even for the young.

Count Bernstorff knew his way around the American press. He also knew what the effect of stories of children deprived of milk, butter, bread and other essential foodstuffs would have on large sections of American public opinion. It was, in fact, in his view, the only way in which to justify in American eyes a return to unrestricted U-boat warfare on the part of Germany. He therefore organized a tour by American journalists, representing all shades of American press opinion, to report on the sufferings inflicted on children in Germany as a result of the British naval blockade.

What Count Bernstorff did not know was that the Berlin War Ministry – no doubt still employing officers whose attitude was not unlike that of the officer dealing with the Nurse Cavell tragedy – was at that moment arranging a carefully organized tour by a different set of American journalists with the object of proving that Britain's naval blockade was having no appreciable effect on Germany's food supplies, in fact that as a result of the blockade the diet of children was more balanced and healthier than it had ever been before.

The accounts by the American journalists based on the tour organized by the Berlin War Ministry appeared first, and Count Bernstorff who knew the story he would have liked to have been put across to the American public to have been nearer the truth, cancelled the tour he was organizing. He did, however, protest against the Berlin War Ministry's action and enquired politely, though pointedly, whether articles by neutral American journalists, suggesting that Germany was suffering no hardship, as a result of the British naval blockade, were not going to make it very difficult to justify the re-imposition by Germany of unrestricted U-boat warfare should that prove necessary – as it did in 1917. The reply he received was that the purpose of the series of articles sponsored by the Berlin War Ministry was not to win sympathy from America but to show the British that they had failed to bring Germany to her knees by starvation.

Count Bernstorff might have pointed out – although he did not do so explicitly – that the target of the articles was America and not Britain. And this argument would have been reinforced when the United States declared war on Germany in April 1917 two months after Germany felt compelled to reimpose unrestricted U-boat warfare in February 1917 in order to try to break the stranglehold of the British naval blockade.

Apart from the establishment of a clear objective and an accurate analysis of the target to be aimed at, there are two other ground rules which a propagandist disregards at his peril. The first is that the message he is trying to get to his target must carry credibility. If it does not, he is not likely to win his target's attention or, should he have won it, to retain it for very long.

Credibility depends on a number of factors. For example, your propagandist may not be talking the right language, as far as his target audience is concerned. His language may be out of date. So may be its pronunciation. William Joyce, more popularly known as Lord Haw-Haw, failed on pronunciation as much as anything else. His carefully cultivated middle- or upper-class English accent undermined any chance of his 'message' getting across to his British target audience. His message in 1939 and 1940 was to play on the social tensions in British society after the hungry 'thirties', but even the most under-privileged of the hungry 'thirties' found a man like Joyce hiding his lower middle class and Irish origins behind a patently assumed upper middle-class accent simply not credible. As a result he became a music-hall joke – Lord Haw-Haw – instead of being a serious political agitator exploiting the social and income divisions within Britain. Whether he would have had any measure of success as a political and divisive social force within British society is doubtful. It was not only his supercilious voice that worked against him, it was his lack of understanding of the basis of social tension in Britain.

Joseph Goebbels, the Nazi Propaganda Minister, had a poor estimate of William Joyce, alias Lord Haw-Haw, but while the Second World War was on he never succeeded in stopping him from broadcasting. Nor did he stop – at least for many months – broadcasts by Radio Free Caledonia or Radio Free Wales. Goebbels may have assumed from newspaper cuttings of the middle and late 1930s and from the reports of Ambassador von Ribbentrop in London, that there were national Scottish and Welsh movements struggling to be free of rule from London, but there was little love lost between Goebbels and Ribbentrop, and it is well known that Goebbels kept only his own press cuttings.

If, therefore, a man as powerful as Goebbels was unable to silence Radio Free Caledonia or Radio Free Wales, it was because the German bureaucratic machine, once launched, could not be stopped in its tracks. As a result, so-called 'black' radio stations like Radio Free Caledonia and Radio Free Wales – 'black' in this context meant stations which pretended to be what they were not, i.e. 'freedom' stations broadcasting from within so-called 'occupied' areas – went on broadcasting at enormous expense in terms of technical and manpower resources.

And all to little or no avail. For the whole effort lacked credibility. Scotland, Wales and for that matter Cornwall, have always had their nationalists. But in 1940 and 1941 not even the most rabid nationalist in any of Britain's Celtic fringes wanted to be told by a German 'freedom' station allegedly operating within its territory what to do about the oppressive English.

The B.B.C. in its broadcasts to occupied Europe and to the enemy in the Second World War, carried conviction to friend and foe alike because it was a pillar of its policy to tell the truth, and to tell the truth irrespective of whether it was good or bad. In December 1941, when the Japanese were sweeping down the Malayan peninsula towards Singapore, the British battleship *Prince of Wales* and the battle cruiser *Renown* were sent up North from Singapore to attempt to stem the Japanese advance by laying down a heavy artillery barrage with their big guns in front of the Japanese. Their mission failed. Both the battleship and the battle cruiser, lacking air-cover, were attacked and sunk by Japanese land-based aircraft. The B.B.C. gave the news of this disaster to the world at once. Some two months later, in February 1942, Singapore and its garrison – 64,000 men, some 20,000 men more than the British lost at Dunkirk in 1940 – had to surrender to the Japanese. Again, the B.B.C. gave the news immediately and straight. By contrast the German radio lost credibility even among its own people by trying to turn disasters into spurious triumphs. For example, when 250,000 Axis troops had to surrender to the Allies in Tunis in May 1943, German radio commented that this in fact strengthened the defence of the 'fortress Europa' because it rid Germany of commitments outside Europe. The argument carried little conviction.

It was by giving the truth – good or bad – that the B.B.C. won and maintained its credibility. And by pursuing this policy consistently the B.B.C. gained an ever-increasing audience as the war continued, even among dedicated opponents of all the B.B.C., Britain and its Allies stood for. The reason why it reached such an amazing credibility rating, was very simple: in times of war – which inevitably means uncertainty – people would rather know how things stand than to be fed news and information they distrust.

Not that every form of psychological warfare or propaganda depends on truth to be accepted as credible or to carry conviction. For example, in November 1950 the U.N. forces dropped leaflets on the Chinese Communists who had by then intervened in the Korean conflict. The purpose of the leaflets which had been prepared and printed many weeks in advance of November 1950, was to convince ordinary Chinese Communist conscript soldiers that the equipment and rations of U.N. forces – no matter what their nationality – were greatly superior to anything the Communist Chinese could issue. No Chinese Communist soldier disputed that point. What destroyed the impact the leaflet might have made, was that the American psychological warriors who prepared the leaflet, chose to show corpses of Chinese soldiers who had died needlessly 'resisting the U.N. Turkish forces'.

Two factors had prompted them to choose the Turks: in the first place the American psychological warriors wanted, with the best of intentions, to play down the overwhelming American contribution to the U.N. effort in Korea, and, secondly, they tried to give credit – and justified credit – to Turkey's contribution to the U.N. war effort in Korea which was out of all proportion to Turkey's size and military power. No doubt entirely laudable objectives, but a lesson for those involved in psychological warfare why not to seek to attain too many objectives at one and the same time. For what had happened between the devising of the leaflet and its distribution was that the main Turkish combat regiment had been virtually pulverized in an attack by vastly superior Chinese Communist forces. It took about three months to reconstitute the Turkish contingent, and the Chinese learnt later to their cost how effectively it had been reconstituted, but unfortunately it was during those three months that the leaflets were dropped on the Chinese Communist lines.

Chinese soldiers, taken prisoner with the leaflets in the pockets of their uniforms, said that they could not believe what the leaflet said since U.S. as well as Chinese news sources had reported the virtual destruction of the Turkish contingent. Their comment, almost without exception, was: 'Why tell us lies?'.

Officials of the U.S. psychological warfare branch concluded in a somewhat disillusioned way that the truth does not always make for credibility. For example, if the caption on the pamphlet showing dead Chinese soldiers had described them as dying needlessly 'resisting U.N. French forces', the pamphlet would probably have been accepted without question and carried credibility – although there were no French forces in Korea, and the pamphlet would in fact have contained a lie.

In other words, credibility and truth do not necessarily march in step in psychological warfare. What is said, written, printed or broadcast in the field, must be credible within the terms in which it is projected; it need not necessarily in every instance be true. In fact, the truth can – as the U.N. Turkish forces caption in Korea demonstrates – in certain circumstances defeat credibility.

The criterion of what carries credibility within the terms in which it is projected is particularly important in the case of 'black' broadcasting stations, i.e. stations which pretend to be what they are not. Within these terms, Radio 'Free Caledonia' or Radio 'Free Wales' carried no conviction. On the other hand, the 'Soldatensender Calais' – which pretended to broadcast from within German-occupied Western Europe for the benefit of the German armed forces stationed between the Northern Cape and the Bay of Biscay – did. Not because a fraction of its audience ever believed, or continued to believe, that it broadcast from within German-occupied Continental Europe or was a genuine German armed Forces Radio Station. The pretence, the lie that it broadcast from within German-occupied Europe, was irrelevant. What was relevant for the German soldiers, sailors and airmen who had to man observation posts from the northernmost part of Norway to the Atlantic frontiers of Spain, was that the 'Soldatensender Calais' seemed to know exactly what was going on inside German-occupied Europe. A pack of U-boats was slipping out of Lorient? How could it be that 'Soldatensender Calais' knew the members of at least three of the pack? As a German U-boat captain whose submarine in fact happened to be among those mentioned by the 'Soldatensender Calais' explained after he and most of his crew had been made to surface by depth charges and taken prisoner: 'The broadcast gave us the eerie feeling that the British Navy knew all about us. As a result, my crew and myself felt that we were in a submerged coffin.'

The credibility of a 'black' station like the 'Soldatensender Calais' depended entirely on the extent of the information supplied by the various intelligence services. And also on knowing how to use such information as was supplied most effectively. For example, military intelligence may have learnt that two or three new German fighter-bomber aircraft may have exploded on take-off, in mid-air in banking to starboard or port too sharply or on landing. 'Soldatensender Calais' exploited these accidents – inevitable in developing any new fighter-bomber aeroplane – to undermine the morale of young, keen Luftwaffe pilots. Moreover, the 'Soldatensender Calais' pointed out how many non-German workers with no interest in Hitler's ultimate victory were employed in making component parts for the new aeroplane. The figures of the non-German workers employed in this way may not have been true, but no one listening to the broadcast had the means of checking. And even if he had tried to check, he would have run the risk of appearing to doubt the superiority of Hitler's Germany, or, worse still, of being branded as a traitor.

In other words, the young Luftwaffe pilot – like the U-boat captain – was a victim of the regime he served. He could not, dare not voice his doubts, and as a consequence a 'black' station like 'Soldatensender Calais' which both of them knew to be 'enemy' stations, gained credibility. 'Enemy' station or not, it spelled out the fears they dared not express to their subordinates, equals or superiors in the armed forces hierarchy – or even to their families, many of whom acquired the habit of listening to the 'Soldatensender Calais' and asking awkward questions of their menfolk, thereby undermining the fighting men's morale still further.

In the case of a so-called 'black' station like the 'Soldatensender Calais' credibility depended not on truth in the sense of trying to convince listeners of being what it was not – that is to say a genuine German armed forces station broadcasting from inside German-occupied Western Europe – but of conveying the impression in its broadcasts that it knew what it was talking about. It did so partly because it always broadcast in the latest jargon used among German soldiers – officers and men. And the jargon included phrases in which fighting soldiers – whatever their rank – came to refer to Nazi officials and their families far behind the front line and far away, in safe and comfortable country retreats, from the heavily bombed industrial cities of Germany's heartland. It also succeeded in carrying conviction – despite the fact that it was known to originate from outside Nazi Germany – that its stories were built on at least a grain of truth or something that might be true or, to put it lower still, something that nobody on the Nazi side could disprove or attempt to disprove without running the risk of giving more publicity and perhaps greater credibility to a story which might otherwise be quickly forgotten. To take, for example, the case of fatal crashes of new fighter-bomber aeroplanes, it is no secret that in their development, all new aeroplanes demand their toll of pilots and accidents. In trying to develop a jet fighter in the last year or so of the Second World War, Nazi propaganda deliberately underplayed the exceptionally heavy accident rate involved, both in men and material. It was known that the casualty rate was very, very high, and it was suspected – with reason as it turned out after the War – that many of the new German jet fighters were sent into active service without being adequately checked.

The B.B.C. reported only those accidents as facts which were substantiated. What was not substantiated, was reported as rumour. 'Soldatensender Calais' used the small basis of factual evidence of fatal accidents to mount a campaign against the development of the new German jet 'planes as being no more than a callous attempt by Hitler and his gang to kill off Germany's bravest and most skilled young pilots; and it did so with the help of German technical Luftwaffe personnel who talked in terms their former Luftwaffe colleagues, technical or otherwise, would understand.

If the B.B.C. had broadcast or attempted to broadcast along the lines of 'Soldatensender Calais', it would have lost credibility – for its credibility depended on reporting the truth and strictly separating truth from rumour and speculation. By contrast, the credibility of the 'Soldatensender Calais' depended not on truth but on how much rumour and speculation its listeners could be expected to swallow. The credibility of each operation depended on the terms within which it was operating.

There is a fourth ground rule to which anyone concerned with psychological warfare must pay the closest attention if his work is to be effective: to get what he wants people to think or do across to them, to organize his means of communication. It is not enough for anyone engaged in psychological warfare to determine his aim, the direction in which to mould people's attitudes and possibly actions, or to define the target he seeks to influence or to ensure that his message carries credibility. All these three factors are essential, but they will be of no avail unless he manages to reach his target, unless his message gets across. Unless it does, all his work will be little more than an academic exercise of no practical value or impact.

For example, in January 1918 President Wilson delivered his famous Fourteen Points Programme speech to Congress. On reading it, every war-weary German soldier and his family at home were bound to ask themselves whether – if these were the terms offered – there was really much sense in continuing the fighting and killing. But they were only going to be able to ask themselves that question if they were given a chance of reading the 'Fourteen Points'. And the chances of their reading President Wilson's 'Fourteen Points' were small.

Of course the neutral press in Europe – in Switzerland, Holland and Scandinavia and to a much lesser and to some extent distorted, degree in Spain and Portugal – carried the text of the 'Fourteen Points'. But the impact of the neutral press on German public opinion and on the German-speaking population of the Austro-Hungarian empire was minimal. Only a tiny portion of the population in either Germany or Austro-Hungary could get hold of neutral newspapers. If the 'Fourteen Points' were to have any effective impact, they had to be spread widely and speedily. Otherwise, the fighting and the killing could by mischance be spread right across a fourth, and perhaps a fifth, winter and spring of war.

The problem, therefore, was one of communication. To have people who could afford either directly or through friends to lay their hands on neutral newspapers reporting President Wilson's 'Fourteen Points' and discreetly discussing, among themselves, the possible implications of President Wilson's speech, was one thing; to have ordinary men and women in Germany and Austro-Hungary who had to bear the heat and burden of the day – and most of the casualties and ever-increasing day-to-day hardships of food rationing and keeping growing children fed – was another.

It was to the second that President Wilson's 'Fourteen Points' had to be got across. Broadcasting did not exist in the First World War. The problem was consequently very practical.

It was resolved with a large measure of success by an American, Edgar Sisson, who was the representative in Russia in 1917 and 1918 of George Creel, President Wilson's special adviser on propaganda in Washington. Although cut off from ordinary news sources, Edgar Sisson, being a highly sensitive political animal, got wind of President Wilson's plans for a speech to Congress, on 8 January 1918, in which the President proposed to outline America's peace aims. The content of the President's speech, he guessed and guessed correctly, could be as important to the belligerent nations of Europe as to the American

Congress. He therefore arranged with his superior in Washington, George Creel, to have the text of the speech telegraphed to him in Russia as quickly as possible. Once he received the text, he realized that his instinct had been right. Here was something that could change people's attitudes and possibly behaviour among the belligerent nations of Europe. But within his own terms of reference as well as for purely practical purposes, the attitudes that mattered most were those of Germany and her Allies. How to get President Wilson's message across where it mattered most at that moment of time?

Sisson tried to put his case to a number of Soviet hierarchs. He got nowhere. Eventually after a few days he managed to get an interview with Lenin. Lenin saw the political impact President Wilson's 'Fourteen Points' could have at once. He also noted President Wilson's condemnation of Imperial Germany's harsh terms in imposing the Treaty of Brest-Litovsk on Russia following Russia's military collapse. He expressed his regret that the United States was not prepared to go one step beyond expressing regret at Germany's harsh terms and formally recognizing the Soviet Union as the official government of Russia, but he did not press the point. The most important issue at the moment, he told Sisson, was to have President Wilson's 'Fourteen Points' as well as the references to Russia's cruel treatment by Germany translated into Russian – for distribution inside Russia 'to give heart to our own people' – and to have it translated into German. Both translations were to be printed on as many pamphlets as possible, and their printing and distribution was to take precedence over any other pamphlets.

As a result of Lenin's orders more than a million pamphlets of President Wilson's 'Fourteen Points' were printed. Tens of thousands were handed to German and Austrian prisoners of war in Russian captivity as they were being repatriated to their native lands under the terms of the Treaty of Brest-Litovsk. Tens of thousands of similar pamphlets were strewn in the paths of the German and Austro-Hungarian armies which were advancing into the Ukraine in early 1918 in order to try and relieve the deteriorating food situation in Central Europe. To ensure that President Wilson's 'Fourteen Points' secured their maximum impact, Lenin had the President's message translated into Czech, Slovak and Hungarian as well as German.

As a result of Edgar Sisson's initiative in revolutionary Russia, the ordinary people of Germany and of her Central European allies knew all about President Wilson's 'Fourteen Points'. But for him the propaganda impact of President Wilson's speech would have been gravely diminished, if not entirely lost.

The development of broadcasting between the two World Wars made the means of communication with your target areas a good deal easier in the Second World War than it had been in the First. But important though broadcasting became in the Second World War, other methods of communication continued to play an important part. Leaflets dropped over or fired into the enemy lines still played an effective part, especially when they were meant to be used as 'passports', so to speak, to surrender. In sustaining home morale, all the belligerents in the Second World War found their domestic newspapers – national or local – the most useful medium of communication. As for spreading rumours inside the enemy camp, an item about, say, the inadequacy of a certain piece of military equipment, planted in a respected neutral newspaper, could cause greater unease and concern in the target area than a similar item disseminated by different means.

Aim and purpose, target, credibility, and means of communication or dissemination – these, broadly speaking, are the four ground rules to which any propagandist, anyone engaged in any psychological warfare operation big or small, must pay the closest attention. But it would be a mistake to assume that these rules can be applied as if they operated in separate watertight compartments, that if a psychological warrior got two out of four right, he might be half successful, if three out of four, his chances of success would improve proportionately, and if four out of four, complete success would be his. On the contrary, these ground rules operate in such a way that one acts and reacts upon the others. They are interlocked, and the relative importance that each plays in every case and at any moment of time varies from operation to operation.

In their English-language broadcasts to the United States, for instance, the Japanese sought to convince the Americans that America had no reason to continue fighting. The United States had no ideas about its war aims. It did not know what it was fighting for. So why go on fighting?

The aim and purpose of this line of propaganda was crystal clear. So was its means of communication and dissemination – short-wave radio transmissions. But it showed a complete lack of understanding of the target audience, a total failure to think oneself into the skin of or interpret the state of mind of the majority of Americans in the weeks and months following Pearl Harbor, when 'Remember Pearl Harbor' was more than adequate for fighting on and no other war aim was needed. This monumental insensitivity on the part of the Japanese propagandists can be put down to not bothering to studying their target. Certainly it robbed their propaganda of all credibility.

By contrast, the broadcasts in 1945, of a U.S. naval officer, Captain Zacharias, who had served as a naval attaché in Tokyo, were textbook examples of psychological warfare. The aim of his broadcasts, which began after Admiral Suzuki became Japanese Prime Minister following the fall of General Tojo, was to try and convince some of the élite ruling Japan that their country had more to gain by ending the war than by continuing it. His target was the group of admirals close to the Emperor whose attitude collectively and singly had been carefully scrutinized and analyzed over the years and was thought to be more in favour of peace, subject to certain conditions, than the war party led by General Tojo. Captain Zacharias' credibility rested on the correct use of the Japanese language, on using phrases and formulations which Japanese naval officers of distinction would use among themselves and on making it clear that he was speaking in the name of the U.S. Government. That his broadcasts were getting across and were having some effect, was borne out by the fact that they eventually evoked a response. Whether they might eventually have led to negotiations, we shall never know. The atom bombs on Hiroshima and Nagasaki brought about Japan's surrender, and the eventual effect that Captain Zacharias' broadcasts might have produced can only be a matter of speculation.

Sometimes propaganda requires a vast organizational structure sustained by immense resources of men and material. At other times, it needs no more than a simple letter. In September 1940, as the Luftwaffe was beginning its battle to win control of the skies above Britain and the U-boats sought to cut Britain's sea supply lines, Franklin D. Roosevelt announced that he was handing 50 old U.S. destroyers over to Britain in exchange for the lease of British bases for the U.S. Navy in the Caribbean. Many Americans were suspicious of their President's move. They saw it as a step that was likely to get their country involved in somebody else's war. A letter from General Pershing, the commander of the American Expeditionary Force in the First World War, a hero and a man respected by Americans of different political views for his honesty and integrity, stilled many of these doubts. It stated in the straightforward, simple language that America expected of the General, that he for one saw the destroyers-for-bases exchange first and foremost as being in the interests of making America safer in an increasingly unsafe world.

Whether the operation is simple or elaborate, expensive or relatively cheap, all propaganda, every form of psychological warfare must observe the ground rules. As the experience of the two World Wars has shown, psychological warriors ignore them at their peril – just as they jeopardize the chances of success in their work by trying to be too clever.

PART ONE # First World War
1914—18

CHAPTER 2   # Seizing the
Propaganda Initiative

General Ludendorff, who was First Quartermaster-General, a position equivalent to Chief of Staff, of all the Imperial German Armies from 1916 to the end of the War, was lavish in his praise of British propaganda when he wrote his *War Memories* after the War. British propaganda, he claimed, had undermined German 'readiness to fight. . . . We no longer battled to the last drop of our blood. Many Germans were no longer willing to die for their country. The shattering of public confidence at home affected our moral readiness to fight. The attack on our home front and on the spirit of our Army was the chief weapon with which the Entente intended to conquer us, after it had lost all hope of a military victory'.

By contrast, his references to German propaganda were scathing. 'Our political aims and decisions, issued to the world as sudden surprises, often seemed to be merely brutal and violent. This could have been skilfully averted by broad and far-sighted propaganda. . . . German propaganda was only kept going with difficulty. In spite of all our efforts, its achievements, in comparison to the magnitude of our task, were inadequate. We produced no real effect. . . .'

The description of British and Allied propaganda as 'the chief weapon with which the Entente intended to conquer' Germany may seem excessively complimentary, especially since it comes from a man whose final offensive against the Allies on the Western Front in 1918 had failed and who felt compelled under the pressure of the Allied counter-offensive to advise his commander-in-chief, Field Marshal von Hindenburg, and the Kaiser to sue for an armistice. Yet there is no doubt that right from the start of the First World War in 1914 British propaganda at home, in enemy countries and among the neutral nations tended on the whole to be more effective than German propaganda. The reason for this was that Britain had to think about propaganda – and think about it seriously – before any of the other belligerent powers were compelled to do so. For by chance rather than by design Britain was the only belligerent power in which there was serious argument on whether to enter the war at all.

Between 28 June 1914, the day on which the heir to the Austrian throne, Archduke Franz Ferdinand and his wife had been assassinated in Sarajevo, and the beginning of August, Austria, Serbia, Imperial Russia, Germany and France had exchanged ultimata, and to show that they meant business had ordered partial or total mobilization. All the major Continental powers had conscription, and the elaborate mechanics of mobilization served to stifle doubts about the justice of going to war even in countries like France, where it might have been possible to express them. Not that there were many, even there, who wanted to. Everywhere the men answered the call to arms, carried along on a popular patriotic wave, convinced of the righteousness of their country's cause.

It was different in England. The Liberal Party, which had been in power since 1905, was predominantly anti-war. So was the Labour Party. Sir Edward Grey, the Foreign Secretary, received no support for his policy of coming to France's aid, from C. P. Scott and C. E. Montague of the *Manchester Guardian*, from the *Daily News,* the *Nation* or the *Economist*. A group of bishops and Oxford and Cambridge scholars, including Gilbert Murray, the great classicist, pleaded for neutrality. Mrs. Fawcett of the National Union of Women's Suffrage Societies addressed a women's protest meeting against war in the Kingsway Hall. Expressing the reservations felt by many in the high summer of 1914, Professor Sir Llewellyn Woodward, then a young graduate at Oxford, asked: 'Why should we interfere in what was primarily an Austro-Russian dispute over areas of political influence in south-eastern Europe? What claim had society upon me for help in getting it out of the political impasse, into which it had blundered? Above all, would I be justified in killing Germans? Killing was murder, whatever the recruiting slogans might say about my King and Country needing me'.

Even inside the British Cabinet the Prime Minister, Mr. Asquith, and Sir Edward Grey, the Foreign Secretary, met opposition from some of their most influential colleagues. Their predicament was aggravated by the fact that not all Cabinet members knew of the secret Anglo-French staff talks which had taken place in 1906 following the establishment of the Entente Cordiale in 1904, or of the Anglo-Russian agreement of 1907 setting up the Triple Entente.

The argument was resolved by the Germans. Their plan of attack against France, drawn up by the late Count von Schlieffen, made it absolutely essential for them to invade Belgium, a country whose neutrality had been specifically guaranteed by the Great Powers, including Germany. The Germans contemptuously swept aside all protests against their breach of Belgian neutrality by dismissing the guarantee as a 'mere scrap of paper'. What was more the Belgians resisted. Here was the issue to unite the nation. Even the doubters – with few exceptions – rallied to the defence of 'brave little Belgium'.

By the end of 1914 nearly 1,900,000 men had enlisted in the Army, including Sir Llewellyn Woodward. By September 1915 the number of volunteers had risen to over 2,250,000.

On 5 August 1914, the day after Britain's ultimatum to Germany demanding the withdrawal of all German troops from Belgium and Luxembourg soil had expired, the Prime Minister, Asquith, declared in the House of Commons that Britain was fighting 'for the principles whose maintenance is vital to the civilized world'. The Government also took the trouble to publish a Blue Book showing the avenues the Government had explored and the stones it had turned in the pursuit of peace. As a result the opposition to the war, or what had in some quarters been merely grudging acceptance of it, evaporated with remarkable speed. The Labour Party and the trade unions swung their weight behind the war effort. Only Ramsay MacDonald clung to his anti-war views and had to resign from the chairmanship of the Parliamentary Labour Party. Mrs. Fawcett urged the members of the National Union of Women's Suffrage Societies to put their organization at the disposal of the nation in its hour of need. C. P. Scott and H. G. Wells came actively to support Britain's war effort. Dr. John Clifford, a former President of the National Free Church Council and leader of the Pacifist opposition to the Boer War who had returned from a conference of the Churches' Peace Alliance in Germany in the last days of peace, had been 'in favour of a rigorous abstention from joining in the war'; he now announced that he had become convinced 'that our Government had done everything that could be done to allay the storm and preserve the peace of the world. Since Germany deliberately and of express purpose and according to long prepared plans, had broken into Belgium, flung to the winds as veriest chaff her solemn treaty obligations, flouted public law, and trampled under foot with ineffable scorn the rights of small nationalities as not even the small dust of the balance', he proclaimed support for this war to be as much a religious duty as ever opposition to the Boer War had been.

The Belgian issue also caused Gilbert Murray, who had signed the plea for British neutrality, to change his mind. It was an agonizing reappraisal, as becomes painfully obvious on reading his explanation entitled *How Can War Ever be Right?* In it he concedes that 'as far as the rights and wrongs of war go, you are simply condemning innocent men, by thousands and thousands to death, or even to mutilation and torture'; yet he goes on to argue that war cannot be judged simply as 'a profit and loss' account, forgetting 'the cardinal fact that in some causes it is better to fight and be broken than to yield peacefully'.

Gilbert Murray was not the only one whose conversion came about after much agonized analysis. The number of those who were opposed to or had doubts about Britain joining the war, may not have been great, but they were to a large extent men and women who were educated and who held positions of influence. As a result Britain was compelled to recognize the fact – before any of the other belligerents – that it had to justify the righteousness of its cause, in short that propaganda was an essential part of the war effort.

The first body which was established to disseminate propaganda for domestic consumption was a private venture, the Central Committee for National Patriotic Organizations. Its Honorary President was the Prime Minister, Asquith, with Lord Rosebery and A. J. Balfour as Vice-Presidents. The Committee invited prominent people to lecture or write on the causes of the war with a view to justifying 'both historically and morally England's position in the struggle'. The Committee secured from six members of the Oxford Faculty of Modern History a publication entitled *Why We Are At War: Great Britain's Case*. This was followed by the so-called 'Oxford Pamphlets', almost a hundred in number, some of which – it must be admitted – were remarkable for their patriotic fervour rather than for their academic objectivity and integrity.

The traumatic experience of the agonizing arguments over whether or not Britain should enter the War, had convinced the Government that propaganda was too important to be left to the Central Committee for National Patriotic Organizations. After all, the arguments had flown back and forth across the table in the Cabinet Room at No. 10 Downing Street. If Asquith and Grey had had difficulty in carrying some of their colleagues with them was there not also a need to convince the British public, the Empire, the neutrals, especially America, of the justice of the British case?! And was this need not likely to continue as the War went on?

The special circumstances surrounding Britain's entry into the War explain why Britain was the first of the belligerent powers to set up an official war propaganda apparatus. A few weeks after the outbreak of the War the Prime Minister, Asquith, and the Foreign Secretary, Grey, asked Mr. Charles F. G. Masterman, then a member of the Cabinet, to organize and take charge of a propaganda bureau for the British

Government. Masterman may at first sight seem a curious choice for such a post. His main interest was social reform. He had had a chequered parliamentary career and had at one time been Financial Secretary to the Treasury. He had, however, shown considerable resourcefulness and resilience during his career, and political friend and foe alike acknowledged his executive ability. Moreover, he was a close personal friend of Asquith and throughout the first two years of the War while Asquith remained at No. 10 Downing Street, he had direct access to the Prime Minister. He also had direct access to Grey at the Foreign Office, an invaluable asset for a propaganda chief in wartime.

At the time of his appointment, Masterman was Chancellor of the Duchy of Lancaster and Chairman of the National Health Joint Committee. The offices of the National Health Committee were then in a building called Wellington House at 8, Buckingham Gate, and Masterman decided to borrow Wellington House for his propaganda bureau and also some of the Health Committee's most brilliant civil servants, as well as recruiting men like the historian Arnold Toynbee of Balliol College, Oxford. Masterman felt it essential to the effectiveness of his department to keep its existence as secret as possible so as to hide the source of any propaganda material which was put out, since if the source were known, recipients might regard it as tainted. The obscure offices of the Health Insurance Committee in his view provided the propaganda department with an ideal cover.

In his insistence on secrecy, Masterman had the full support of Asquith and Grey. All who were in the know carefully avoided mentioning the subject, and questions in the House of Commons were either met with evasive answers or forthright refusals to explain in any detail. All that Parliament was told was that their former colleague Masterman was receiving an annual compensation of £1,200 from Secret Service funds, that his chief assistant was a former journalist named G. H. Muir, and that his work was very important. More than these tantalizing morsels of virtual non-information neither Asquith nor Grey were willing to give.

The broad objectives Masterman set his organization were to present the righteousness of Britain at home and abroad, to mobilize hatred against the enemy, to preserve the friendship of allies, to win the friendship and, if possible, to procure the co-operation of neutrals – particularly America – and to demoralize the enemy.

Masterman set out to achieve these objectives by a wide variety of activities in different fields. Wellington House was to produce, translate and distribute books, pamphlets, Government publications and speeches, dealing with the War, its origins, its history and all the varied and difficult questions which arose during its development. It was to assist in the placing of articles and interviews designed to influence opinion in the world's newspapers and magazines, especially in America. It was to ensure the wide distribution of appropriate pictorial matter, cartoons, pictures, drawings and photographs for insertion in newspapers and periodicals and for exhibition. It was to produce and distribute films, maps, diagrams, posters, lantern slides, lectures, picture postcards and all other means of miscellaneous propaganda. Furthermore it was to help provide information and facilities to the London correspondents of neutral, especially American, papers on the sound assumption that reports, particularly reports favourable to Britain, were likely to prove more effective in a neutral country than anything produced by a Briton. Certain members of the staff of Wellington House were also to enter into personal correspondence with influential people abroad, especially in America, and to arrange for visits by distinguished neutrals, especially Americans, to Britain and for visits by distinguished Britons to neutral countries, especially the United States.

From the moment Masterman took up his propaganda duties at Wellington House he showed a keen awareness of the overriding importance of winning American support for Britain and her allies. Though by far the most important propaganda organization in Britain during the First World War, Wellington House was not alone in the field. The War Office, the Admiralty, the Home Office, the Foreign Office and General Headquarters in France all had sections or at least groups of individuals who concerned themselves with some aspect of propaganda, and Masterman spent much of his time and energy in trying to co-ordinate these diverse efforts. There was one aspect of propaganda for which he was not responsible, and that was recruitment. This was the responsibility of the Parliamentary Recruiting Committee, which operated somewhat as a do-it-yourself propaganda organization. It commissioned well over a hundred recruiting posters, all of them of little or no merit with one brilliant exception – Alfred Leete's 'Kitchener'. It made the War Minister immortal, and it seems to be generally agreed that his accusing finger directed many a wavering young man towards the nearest recruiting office. Certainly it is one of the most outstanding posters ever produced for any purpose anywhere in the world.

Masterman and his colleagues at Wellington House realized early on in their work that if they were to be successful in attaining their objectives, they had to know what the enemy was saying about himself and the Allies, what the neutral press was saying, what the psychological mood and temper in Allied, neutral and enemy countries was and how and if it varied between different sections of society. Such intelligence was essential, in Masterman's view, in formulating the content, aim and method of particular propaganda. Indeed, as the War progressed, the Intelligence section became one of Wellington House's most important parts and was one of the main reasons for its effectiveness.

Not that its effectiveness saved it from being buffeted by the political upheavals, personal feuds and inter-departmental rivalries during the four years of the First World War. It saw many changes.

By the autumn the great Departments of State – the War Office, the Admiralty, the Colonial and Home Offices – had become aware of the fact of how powerful a weapon propaganda could be and were all engaged in expanding their own separate propaganda departments. The War Office and General Headquarters in France, in particular, were inclined to dismiss Wellington House as insignificant and inadequate. There was considerable danger that Masterman might be compelled to spend all his time in inter-departmental warfare instead of concentrating his energies on the job with which he had been entrusted. To avoid that danger, Lord Newton was put at the head of Foreign Office propaganda. It was hoped that the prestige of a great name would act as a shield for Masterman and bring about a period of harmonious co-operation between Wellington House and the propaganda efforts of the other Departments of State.

The arrangement under Lord Newton did not last long. In December 1916 Asquith fell from power, and Masterman lost his closest political and personal friend at the very centre of power. Although he continued at Wellington House, his influence within the British propaganda apparatus waned rapidly.

At the beginning of 1917 the new Prime Minister, Lloyd George, sought to consolidate and centralize Britain's propaganda work by forming a Department of Information. The new Director of the Department was Colonel John Buchan, the author of biographies of Raleigh, Scott and Cromwell and of thrillers like *Thirty-nine Steps, Greenmantle* and *The Three Hostages*, and who in 1935 was, as Lord Tweedsmuir, to become Governor-General of Canada.

The new Department was divided into four divisions: (1) Masterman's Wellington House; (2) a cinema department under Masterman's old lieutenant, G. H. Muir; (3) a Political Intelligence department; and (4) the News department. In short, the functions which Wellington House had carried out up to then, were not basically changed. They were re-distributed within the new Department; there was considerable

expansion and of course there was a new boss, Col. Buchan. Buchan –
like his predecessors in propaganda work – was given direct access to
the War Cabinet. He was also given an Advisory Committee composed, among others, of Lord Northcliffe, C. P. Scott and, a little later,
Lord Beaverbrook.

The inter-ministerial jealousies, however, showed no sign of abating,
and in the early autumn of 1917 Sir Edward Carson, the brilliant
advocate and leader of the Ulster Unionists, was appointed as a sort of
propaganda generalissimo over Buchan. Carson was at the time a
member of the War Cabinet, and it was felt that his authority,
combined with his drive and forceful personality, would succeed in
smoothing out rivalries and jealousies.

Carson, alas, proved no more successful in this than his predecessors,
and in February 1918 Lloyd George, tired of continuing newspaper
criticism of British propaganda, decided on a final re-organization. He
abolished the Department of Information and set up in its place a
Ministry of Information. As Minister he appointed Lord Beaverbrook,
one of Britain's most powerful newspaper proprietors, and as Director
of the most spectacular branch of the Ministry, i.e. the organization of
propaganda in enemy countries, he appointed another equally powerful
newspaper proprietor, Lord Northcliffe. In strict hierarchically bureaucratic terms Lord Beaverbrook was Lord Northcliffe's superior, but
neither press lord was willing to make an issue of the point, and Lloyd
George, who felt that this arrangement – however untidy it may have
appeared on paper – effectively silenced his two most vociferous critics,
gave both men the right of direct access to the War Cabinet. Moreover,
Lord Beaverbrook made no attempt to have the organization of propaganda in enemy countries move their offices to the Ministry of Information. He let it set up its own separate headquarters in Crewe House,
tne town residence of the Marquis of Crewe in Mayfair.

In any case Lord Beaverbrook was far too busy drastically reorganizing the British propaganda apparatus to bother much about the special-
ized work done at Crewe House. One of his sweeping changes was the
abolition of Wellington House, although Masterman's services were
retained as 'Director of the Literature Department at the Ministry of
Information' and many of Wellington House's other functions were re-
distributed.

No wonder these numerous chops and changes in the British propaganda set-up caused a sceptical Member of Parliament to remark with
the merest touch of cynicism in the House of Commons in the summer
of 1918: 'We started with Mr. Masterman working practically by
himself at Wellington House, then came Colonel Buchan who was put

over him, and then we had Lord Beaverbrook who was put over Colonel Buchan. We have all three of them now'.

The important point, however, was that the special circumstances which surrounded Britain's entry into the War led to the establishment of a propaganda apparatus at the beginning of the War and no matter how often that apparatus was tinkered with, there was throughout the War an organization equipped to disseminate propaganda. Moreover the special circumstances surrounding Britain's entry gave Britain not only an apparatus but also a cause for the apparatus to spread among its own people and the world at large – the cause of 'brave little Belgium', of small nations at the mercy of ruthless military giants, of the sanctity of treaties, of people to live in freedom as they chose, of humanity, of the ideals of democracy and liberty.

Imperial Germany, by contrast, began the War with nothing which could even vaguely be depicted as resembling a co-ordinated propaganda effort. Nor, if it had had the necessary apparatus, would there have been a cause to disseminate. The Germans in August 1914 were so swept away by their own enthusiasm, so certain of the self-evident justice of their cause that they did not feel the need to explain. Private citizens tried to fill the gap left by their Government, but their efforts rested on a completely false estimate of the attitude of the people they were addressing outside Germany. A Wagner Culture Committee was formed, but a call stressing the necessity to fight for and defend German *Kultur* scarcely had the universal appeal of a call to defend human dignity and the right of small nations. Worse still, scores of German scholars took it upon themselves to write to their foreign friends and to explain the German cause, even to the extent of finding excuses for the violation of Belgian neutrality, an aspect of the War in 1914 which troubled many Germans. Professor Lamprecht, who made a special study of propaganda in the First World War in Germany, described the effect these letters produced when he wrote contempt- uously of the educated man who 'obtained the largest possible goose quill, and wrote to all his foreign friends, telling them that they did not realize what splendid fellows the Germans were, and not infrequently adding that, in many cases, their conduct required some excuses . . . the consequences were gruesome'.

By the time the Imperial German authorities woke up to the fact that an argument was going on in the world at large on the question of which of the belligerents was responsible for the War, that the answer to that question was shaping attitudes and moulding emotions to the two sides and that consequently Germany's case must not be allowed to go by default, they discovered that the Royal Navy had cut the cables between Germany and the United States. It had happened on 15 August 1914. Germany found itself deprived of one of the four main requirements for a successful propaganda operation, the means to communicate. She was shut off from the world's most important neutral country at the moment at which American public opinion concerning war guilt was being formed. There was no point in sending cables from The Hague or Amsterdam because the cables from Holland were also controlled from London.

Although, with the inauguration of a trans-Atlantic wireless service, the Germans were eventually able to reach America with their side of the story, the British and Allied version had had plenty of time to sink in. As a U.S. State Department communication put it: 'The first publication is that which is formative of public opinion and which affects public emotion.'

As a result of setting up a propaganda organization in the first weeks of the War and so having an instrument to disseminate its cause and of gaining a valuable time advantage in influencing American opinion by the cutting of the cables between Germany and North America, Britain seized the propaganda initiative from Germany. It was an initiative which Germany was not able to wrest from Britain in the next four years.

Germany's efforts to catch up were not aided by her inability to devise a propaganda organization. For a nation which had a reputation for efficiency, her action, or rather lack of action, was truly astounding. The only formal co-operation was a Press Conference at which representatives of the War Ministry, the General Staff, the Navy, the Ministry of the Interior, the Foreign Ministry, the Colonial Office, the Ministry of Finance, the District Military Authorities, the Post Office and eventually the Food Ministry met a committee of journalists, chosen to speak for the Press as a whole, two or three times a week. Each Ministry took the chair in turn at these meetings. These Press Conferences were at best little more than channels for official hand-outs and at worst forums at which officialdom had to submit to questions on how the war was going, when final victory could be expected, whether food rations were likely to go up or down; they provided a setting in which officialdom was as often as not almost automatically forced on to the defensive. They were certainly not instruments for a dynamic propaganda drive.

The German Foreign Office, which should have been alive to the importance of winning goodwill and support for Germany among the neutrals, seems to have made no preparations at all for putting over Germany's case. It did not even try to make up for lost time, as reports flooded in from its embassies abroad concerning the argument about who was responsible for the War. Like the majority of Germans, the officials in the Wilhelmstrasse appeared to feel that the rightness of Germany's cause was self-evident and needed no justification. In any case, was the War not going to be over by Christmas?

The check of the German advance on the Marne brought a rude awakening, and with the prospects of an early peace receding, the Foreign Office set up a special *Zentralstelle für Auslandsdienst* which turned out a vast amount of propaganda material.

But what this *Zentralstelle* produced was at no time properly co-ordinated with what other official bodies turned out. And the most important of these other bodies was the Army. At the outbreak of the War the Army had but a single official who had contact with the Press. But the military authorities even developed an extensive Press Service to report military operations, to edit the Field Press, to control what domestic German or Austrian papers were allowed to reach the soldier at the front, and to carry on propaganda against the enemy.

Since there was no co-operation, the propaganda put out by the military and the civil authorities was frequently conflicting, and as the War wore on, the conflict between the military and the civil authorities became increasingly more pronounced. German G.H.Q. made no secret of its disgust for the Chancellor, Bethmann-Hollweg, when he held out an olive branch in 1916, and when a peace resolution was moved in the Reichstag in 1917, Ludendorff, the Chief of Staff, gave – or as G.H.Q. put it – 'granted' an interview to the Berlin Press in which he made it absolutely clear that he had no patience with talk about peace; what the fighting man at the front, who had borne and was bearing so many sacrifices, deserved, was a victorious peace of dictation. Naturally the Centre and the Left denounced Ludendorff's interview as an inexcusable interference by the military in politics. The military, which had begun to see evidence of the impact of Allied propaganda, retaliated by demanding – as it had on several previous occasions – the establishment of a Ministry of Propaganda, but the Chancellor, who had no wish to be caught in a continual cross-fire between the Reichstag and G.H.Q. as to how German propaganda ought to be conducted, prevaricated. And so German propaganda continued to speak with different voices until the end of the War. The rift between the military and the civilian authorities widened still further when Ludendorff established a special Press Service called the *Deutsche Kriegsnachrichten* which aimed at reaching the German home public directly over the heads of the civil Government and Parliament and the domestic Press. Its aim was patriotic stimulation of the home population, and despite fierce opposition to it from the domestic Press it did surprisingly well at least in the early months of its existence.

Such an unco-ordinated, ramshackle set-up had little or no chance of snatching the propaganda initiative from Britain. There was bound to be confusion and conflict on two of the essential requirements for a successful propaganda drive: the message to be got across, and the target at which the message was to be aimed for maximum possible effect. The conflict between the military and the civil authorities as well as the political parties inside Germany made confusion worse, and Ludendorff's arbitrary and peremptory intervention did nothing to improve Germany's chances of making an impact.

These divisions inside Imperial Germany lessened the impact of German propaganda from the beginning of hostilities. Some civilian German agencies like the German professors before them, stressed the superiority of German *Kultur,* but this was hardly a theme to fire the imagination, as was H. G. Wells' famous phrase that this was 'the war to end war' because it was a war to defend humanity everywhere. Again, the proud claim that Germany was in honour-bound to come to the aid of its Austro-Hungarian ally did not compare too favourably with Britain's claim that it was rallying to the defence of 'brave little Belgium'. As the War wore on, even German soldiers who were sent to stiffen the Austro-Hungarian front i.e. in Eastern and Southern Europe began to wonder why, because of the promises exchanged between the German and Austrian emperors, they should be sent to shore up the ancient Habsburg empire, which was plainly declining rapidly and which was not even predominantly German. The argument that the French intended to invade Germany through Belgium, with Belgian connivance of course, carried little conviction and was soon abandoned – not least because the German military authorities made no bones about the German invasion of Belgium being a prerequisite to victory which would justify the means.

Indeed, Germany's propaganda efforts tended to cancel each other out because of the conflicts between the military and civil authorities. And the fact that the military in their internal conflicts with the civilian authorities usually tended to come out on top, served only to undermine the effect of German propaganda still further. Most of the world, after all, was civilian in attitudes, reactions and emotions, although many men in many lands became soldiers for the duration of the War, but it simply could not comprehend the paroxysms of rage which the activities of Belgian *franc-tireurs* – civilian snipers who shot at German troops from doorways and rooftops – provoked among the

German military. Americans, especially, were incapable of understanding what the Germans were getting at. After all, it was Germany which had violated Belgian neutrality, so why complain? These so-called *franc-tireurs* were defending their country and their families in the only way they knew how. Had Americans not behaved in exactly the same way during the War of Independence against Britain? British propaganda reminded the Americans of this and suggested that these men were heroes and not criminals.

Nor did themes like the 'encirclement' of Germany or Britain's envy of Germany's growing naval strength and industrial power win many friends and influence people. They sounded like excuses for German behaviour and seemed to be concerned only with the fate of Germany and her people and not – as British propaganda tried to be – with the fate of mankind as a whole. German propaganda appeared to suggest that its country's rulers and people were fighting the War only to right real or imagined injustices and slights suffered by the Fatherland; British propaganda sought to convey the impression that for its rulers and people the War was a crusade for the whole of mankind. Moreover, British propaganda, with very few exceptions, did not allow the military to dictate its themes. Admittedly, Czarist Russia was Britain's ally for purely military reasons, but purely military reasons were not sufficient to induce those in charge of British propaganda even to attempt to justify Czarism. This was due not only to the fact the Czarist regime was anathema to a social reformer like Masterman and liberal-minded men throughout Britain and France, but also because those in charge of the propaganda efforts which were being made to win American support for the Allied cause, knew that the alliance with Czarist Russia was a considerable liability to Britain's and France's case in the view of many Americans.

The French, too, in their propaganda stuck to themes that were of universal appeal. Of course they stressed to their own population that they wanted back Alsace and Lorraine which they had had to cede to Germany in 1871, and there were some who advocated the incorporation into France of all the territories left of the Rhine and perhaps beyond. Of course they stressed the frightfulness of the 'Boche' both at home and abroad, but by and large the propaganda put out by their established diplomatic, military and naval agencies, supported by the newly created *Maison de la Presse* which had agents attached to all the French diplomatic and trade missions abroad, was that France was fighting for democracy, for the lofty principles of the French Revolution, for the spirit of La Fayette.

Czarist Russia had no real propaganda apparatus or message for either its own people or the world at large. For the outside world it clothed itself in the mantle of protector of all Slavs – which even to her Allies sounded surprisingly like Russian expansionism under another name and was as far as possible played down. For its own people the rule adopted by the Czarist Government was the very simple one that the people were to be kept in complete ignorance of any reverses.

When the United States entered the War, in April 1917, President Wilson at once appointed a Committee on Public Information to deal with propaganda both at home and abroad. The Committee consisted of the Secretary of State and the Secretaries for the Army and the Navy. Its Chairman was George Creel.

The Committee had the advantage of being backed by the prestige of three of the most important and powerful Departments in the Federal Government. It was also fortunate in its chairman, George Creel, who became, in fact, if not in name, Secretary for Information and Propaganda. He was a newspaper editor who had on one or two occasions been critical of some aspects of Allied propaganda in America while the United States were still neutral – a circumstance which gave him greater influence with some members of Congress than he might otherwise have enjoyed. And he was a close friend of President Wilson and had ready access to the White House – which made it possible to integrate the formulation of policy and propaganda very closely at every stage. Moreover, he was a man of great resourcefulness, determined, hard-working and full of verve and energy.

By the time the United States entered the War, Creel had no neutrals to worry about. He could concentrate all his energies on getting across the great universal humanitarian ideas of a Wilsonian peace, to make the world safe for democracy. Unless Ludendorff could force a victory in France before the Americans could be brought across the Atlantic in sufficient numbers, Germany and the powers associated with her would have lost their last chance of seizing the propaganda initiative from the Allies.

CHAPTER 3

Atrocity Propaganda

A belligerent nation once having convinced itself, its allies, and wherever possible, the neutral world that it was forced into war in the first place by an unscrupulous enemy despite its own noble, human and peaceful instincts, finds itself compelled sooner or later into searching for reasons for continuing the fight. The prospect of eventual victory is of course the strongest motive, and so long as that prospect shines strongly and brightly, no nation – once it is launched on war – is going to lay down its arms, especially as the enemy, inspired by similar motives, is unlikely to offer a cessation of hostilities by negotiation.

Having sought to pin war-guilt and all the moral condemnation this entails on the enemy, the next step is virtually inescapable: to make the enemy appear inhuman, degraded, foul, incapable of any humane or decent instinct. Until the faith of one side or the other in ultimate victory begins to crack, efforts to place responsibility for the outbreak of war on the other side is bound to be followed by propaganda about the atrocious way the other side is behaving.

The progression is inevitable. Modern wars, which involve every stratum of society and not merely a limited number of professional soldiers, cannot be fought unless the peoples drawn into the war feel that they are fighting for a righteous cause, and they will not continue fighting unless their enemy is presented to them as the incarnation of evil itself.

That at least is what the propagandists in the First World War felt on both sides. They used atrocity propaganda deliberately with two targets in mind: in the first place they aimed it at their own people with the purpose of stirring up hatred against the enemy, and, secondly, they aimed it at neutral powers, especially America, hoping to win support – active if possible – for their cause.

There is nothing lofty or morally uplifting about atrocity propaganda, and many soldiers have given it as their opinion that war itself is an atrocity, and that it is misleading to imagine that war can be fought cleanly. Napoleon said that it is impossible to make war with rose water. General Sheridan, one of Abraham Lincoln's successful generals in the American Civil War, told Bismarck at the Prussian Field Headquarters in the Franco-Prussian War of 1870–71, that war is best got over as quickly as possible; 'the people must be left nothing to weep with but their eyes over the war'. And Lord Kitchener, Britain's War Minister in the first years of the 1914 war, remarked when a certain 'outrage' was being discussed in his presence: 'What is the good of discussing that incident? All war is an outrage.'

Yet in a War into which everyone is drawn, the public needs a devil it can hate. Not one of the belligerent countries failed to indulge in this kind of propaganda, and not one was slow in producing 'facts' to back up its tales of atrocity. It is to their credit that a handful of scholars and journalists in all the belligerent countries later investigated matters a little more closely and usually – not always – discovered that the 'facts' were rarely such as would be accepted in a court of law.

It would, however, be idle to deny that atrocities were not committed in the First World War by all the belligerents – though, as Signor Nitti, Italy's post-war Prime Minister put it – 'to different degrees'. In the emotionally charged atmosphere of the War the propagandists were not interested in matters of degree. In his notorious Hymn of Hate Lissaner wrote this stanza:

> *Hate by water and hate by land;*
> *Hate of the heart and hate of the hand;*
> *We love as one, we hate as one;*
> *We have but one foe alone – England.*

French and British propaganda may not have been so explicit, but the themes that were persistently pursued were broadly similar among all the belligerents. Prisoners of war were ill-treated, under-fed and tortured; women were raped and had their breasts cut off; boys had their right hands severed either to satisfy the enemy soldier's natural sadism or to prevent the victim from doing military service when the time came; men were mutilated and murdered; babies were crucified; civilians were burnt alive; places of worship were deliberately desecrated and works of art destroyed. The enemy was bestial and beyond reform. If mankind was to be saved, he had to be destroyed.

The Germans were not surprised that the French accused them of unspeakable bestialities, that atrocity stories of the 1870–71 War were revived (like the Uhlan killing a four-year-old French boy with his lance after the lad allegedly had pointed his wooden toy pistol at him; or that French scientists had apparently discovered that the Teutonic brain was different from that of other races in that it lacked the part which allowed men to make comparisons and form a balanced judgement of the world in which they lived. It was the absence of this part in the Teutonic brain, this so-called scientific theory suggested, which made it possible for German soldiers to wear their curiously shaped spiked helmets. Nor were the Germans unduly upset when they learnt that the French had opened a permanent 'atrocity museum' in Paris in 1916 under the patronage of the then French Minister of Education. After all, Germany and France had been traditional foes for well over a century, and neither side expected anything but the worst of the other.

What did surprise the Germans and caught them unawares was the volume and the intensity of the anti-German hate propaganda which emanated from Britain from the first weeks of the War. What they under-estimated in Britain was the impact of the violation of Belgian neutrality, especially on so-called liberal opinion. A country which could invade a neighbour it had solemnly pledged not to invade and excused its actions by dismissing its written promise as a mere scrap of paper, was capable of anything. And whatever appeared to support its fiendishness was readily acceptable without necessarily looking too meticulously at the evidence on which atrocity stories were based.

The sense of outrage felt by many in Britain was fed by the more than 180,000 Belgians who sought refuge in Britain in the late summer and autumn of 1914. Admittedly, some of the Belgian refugees may have told their interrogators anything they thought their British hosts wanted to hear. There is always a temptation to do so in such circumstances. But such motives cannot be imputed to the majority of Belgian refugees. They were in a state of shock. They had been driven from their native land. In some cases they had seen their homes burnt down. In others they had been caught in the cross-fire of German and Allied forces. In others still they had been separated from members of their family of whom they had received no news and could expect none. Often they had lost all they possessed. Life over the previous weeks had been one long atrocity for them.

Such was the climate of opinion in Britain when a remark the Kaiser had made some 15 years before was dug up. The Kaiser was probably one of the most compulsive public speakers of his time. He could not resist the slightest pretext for delivering a speech or sending off a message. The occasion was the Boxer Rebellion in China. The German Minister had been murdered, the diplomatic quarter in Peking was under siege, and Germany, Britain, France and the United States sent a combined force to rescue their beleaguered diplomats and restore order in the Imperial Chinese capital. The Kaiser seized the opportunity to send the German contingent in the joint force a message admonishing it to behave like the Huns of old.

And so was born the spike-helmeted, bull-necked 'Hun' of thousands upon thousands of First World War cartoons, depicted as bayonetting babies and women and torturing civilians and wounded soldiers. The Kaiser of course became Attila the Hun. The only protest against calling the Germans 'Huns' came from a group of English scholars who argued that identifying the Germans with the Huns was an insult to the Mongols living in Central Asia.

The Germans realized too late the intensity of the atrocity campaign against them and the impact it was having, especially in the neutral world. They were appalled. And the shock was felt not only in official quarters. Eminent workers, scholars and scientists were aghast that such crimes could be alleged against their nation. Ninety-three of them issued a manifesto called *'Es ist nicht wahr!'* 'It is not true!' As its title implies, it consisted mainly of a series of sweeping denials. It was not true, that Germany caused the war, that Germany had lightly violated Belgian neutrality, that any Belgian was hurt 'except for the most serious reasons of defence'. The manifesto also counter-charged that the Belgian population ambushed German soldiers, mutilated the wounded and murdered doctors at their work. It was even more specific in its allegations of atrocities committed against the Germans in Eastern Europe: 'In the East ... the blood of our women and children drenches the earth and in the West dum-dum bullets tear apart the breasts of our warriors. Those who ally themselves with Russians and Serbs, and incite Mongols and Negroes on to fight the white race, have the least right to call themselves defenders of civilization.'

The manifesto was signed, among others, by Professor Karl Lamprecht, who was after the War to write very scathingly of the attempts of German intellectuals to justify their country's case, world-famous playwrights like Gerhart Hauptmann, theatrical producers like Max Reinhardt and scientists like Wilhelm Röntgen and Max Planck.

What effect did this manifesto have? The German public was delighted. But then the German public in late 1914 was not prepared to believe British or Allied atrocity propaganda anyway. So among the Central Powers the manifesto served no purpose other than confirming the general public in views which it showed no signs of abandoning no matter what anyone said.

As regards the rest of the world – Allied and neutral – its effect was positively harmful. The manifesto was undoubtedly a *cri-de-coeur* by honourable and distinguished men, but it ignored one of the basic 'don'ts' of propaganda – that is to say, to deny an enemy's allegation, be it true or false, thereby giving that allegation fresh and unsolicited publicity. The denial puts its authors on the defensive, and what sticks after a short while, is not the denial but the original allegation. As an American war correspondent, Frederick William Wyle, who was personally far from convinced about the truth of alleged German atrocities in Belgium, put it, 'among the real atrocities of war for which the Germans are responsible, is their mania for self-defensive manifestoes'.

In the British House of Commons, the German professional manifesto served to provoke questions about how quickly the Committee appointed to look into Alleged German Outrages in Belgium would be able to submit its report. This was a Committee set up under the chairmanship of Lord Bryce, who had been Professor of Civil Law at Oxford University, a Liberal Member of Parliament and who had served as British ambassador in Washington from 1907 to 1913. He was a close personal friend of President Wilson, and as an eminent scholar had been awarded an impressive number of honorary doctorates by German universities before the War. His associates on the Committee included H. A. L. Fisher, the distinguished historian, and Sir Frederick Pollock, an authority on constitutional law.

Although the House of Commons was eager to have this Committee's findings as quickly as possible, such a body was not to be hurried. In fact, it did not produce its report until five days after the *Lusitania* had been sunk without warning by a German submarine with the loss of 1,198 lives, 128 of them American, on 7 May 1915.

The timing of the publication of the Bryce Report could not have come at a worse time for Germany – a circumstance which provoked a good deal of caustic comment from German sources at the time during the War and from American scholars after the War. Research has in fact revealed since that the Bryce Report was in the hands of the printers well before the *Lusitania* sinking and that its publication depended entirely on the vagaries of war-time printing schedules.

Far more important than the timing of the publication of the Bryce Report is its contents. In its introduction the Report pointedly stated the case for scepticism: 'It is natural that much of the evidence given, especially by the Belgian witnesses, may not be due to excitement and over-strained emotions, and whether, apart from deliberate falsehood, people who mean to speak the truth may not in a more or less hysterical condition have been imagining themselves to have seen things which they said they saw.'

Lord Bryce and his colleagues asserted that they had carefully omitted merely hearsay evidence. Yet atrocity stories of rapes, mutilations and tortures were corroborated by the simple method of passing from mouth to mouth among the Belgians in London and France and unnamed British officials and soldiers. Moreover, the Committee did not itself examine any of the witnesses. The examination was conducted on its behalf by some 20 English barristers. Yet the Report gave no indication, where the witnesses were Belgian, of the language in which the examination was conducted, and it seems not unreasonable to ask how many members of Bryce Committee's investigation team spoke fluent French and Flemish.

Lord Bryce and H. A. L. Fisher both stated after the War that they stood by the broad findings of their Report, but it is the vast amount of detailed atrocity stories contained in the voluminous appendices to the Report that attracted and held world attention. Almost 60 years after the event it is probably not unfair to comment that the authors of the Bryce Report were not entirely unaffected by the hysteria of the times.

More important, within the context of this book, is to assess the impact of the Bryce Report. It helped to confirm Allied and a great section of neutral opinion that the 'Hun' was an inferior species of the human race. And it put the Germans once more on the defensive. Once again they brought forward the argument that the Belgians had ill-treated German citizens who lived in Belgium at the time of the German invasion – an argument that won them little sympathy in any quarter. Once again they claimed their soldiers had been shot by *franctireurs,* by Belgian civilians hiding in ambush.

Equally defensive were the answers to American protests against the sinking of the *Lusitania.* The ship had been carrying arms and munitions. Furthermore, it had been armed, and even if it should prove not to have been, its captain and officers had instructions to use the ship's superior speed to ram any U-boat that might attack them. U-boats therefore could not surface and give adequate prior warning to the *Lusitania.* They had no choice but to treat the ship as a 'man o' war'. In view of the loss of well over 1,000 lives their line of argument sounded tortuously legalistic and callously inhuman. It took two more notes of protest from President Wilson before the U.S. Government was informed that German U-boat commanders had been instructed not to sink liners without warning and without providing for the safety of passengers.

Defensiveness, too, marked German reaction to charges of launching a new kind of warfare, contrary to international law, in using gas for the first time on the Western Front in the spring of 1915. The French had used gas before them, they claimed, and in any case gas warfare was not against the tenets of international law. Again, the German case was obscured in a tangle of virtually incomprehensible legal argument. And just as indignation over gas warfare began to die down, the Germans found themselves compelled to deny a story about a half-gassed Canadian soldier having been crucified by his German captors – a story which was eventually found to have as little substance to it as many of the atrocity stories in the Bryce Report.

How was it that the Germans allowed themselves to be manoeuvred into this defensive posture on atrocity propaganda? Why, for example, did they not exploit the invasion of East Prussia by the Russians in the early weeks of the War? After all, as the 93 German professors, scientists, artists and writers pointed out in their manifesto protesting against Allied atrocity propaganda, every German was convinced of Russian and Mongol atrocities, and as far as the rest of the world was concerned, one has only to mention Cossacks and Kalmucks invading a town and leave the rest to the imagination. Why, then, were the Germans so reticent in the early days of the War? The answer is partly that the military had too much say in propaganda matters. The General Staff had no wish to let its own people or the world at large know how dangerous the situation was in East Prussia. Moreover, what was happening showed up plainly that Bismarck had been right when he had warned Germany against ever fighting a war on two fronts. The German General Staff did not wish to be shown to have ignored so distinguished a warning and as a result to have exposed its fallibility.

Once Hindenburg and Ludendorff had defeated the Russians at Tannenberg, the attitude changed. There was no crime, no bestiality, no enormity of which the Russians had not shown themselves capable. There were cartoons of Russian soldiers with human fingers dangling from their belts and murmuring to themselves: 'German finger, Austrian finger, Hungarian finger – I don't care which, so long as there's a ring on it.'

The only trouble was that by the time the Germans and later the Austrians got around to disseminating atrocity propaganda, the world's press was full of the atrocities alleged to have been committed by the Germans in Belgium. Moreover, the German cables to important neutrals like America had as we have seen, been cut and the Central Powers had difficulty in getting their stories across.

Of course their atrocity propaganda had a tremendous impact on their own population, and they succeeded in stirring up the fighting mood of their own troops to a very high pitch of excitement on the Western Front by stressing the atrocities Senegalese, Moroccan, Zouave or Indian troops would commit if they fell into their hands.

5. Europe 1914. An accurate assessment by the Germans, no doubt, but one unlikely to win friends.

6. A German cartoon meant for the home public, showing the distorted coats-of-arms of Imperial Germany's enemies.

7. This British leaflet aimed at dividing rulers and governed by contrasting the resplendent Kaiser and his sons with their starving subjects.

8. The Western Front. Preparing balloons to drop leaflets on and behind German lines.

The Germans called the British naval blockade an 'atrocity', but it did not produce the powerful emotional impact of 'unrestricted submarine warfare' because the Germans were reluctant to admit the success the blockade was having.

All the while anti-German atrocity propaganda continued. The Allies published thousands of cartoons. They had the good fortune of having on their side the outstanding cartoonist of the period, Louis Raemackers of the Amsterdam *Telegraaf,* who though Dutch, was more mordantly anti-German than many British cartoonists and whose work far surpassed in bite, viciousness and acid wit anything any German cartoonist could produce. His work was syndicated all over the world.

When the Turks massacred over a million Armenians in the autumn of 1915 and deported another few hundred thousand to the North Syrian desert where most of them starved, Britain's propagandists suggested that it was the Germans who had goaded the Turks into this atrocity. And once again Germany found itself on the defensive.

The execution of Nurse Edith Cavell, as we have seen, raised anti-German feeling to a degree as high as, if not higher than, the alleged Belgian atrocities or the sinking of the *Lusitania.* The crying of children and the whimpering of women, it was supposed to have been proved by scientists, had a special appeal to the Teutonic spirit. Not infrequently the propaganda overreached itself, and the material used in support of some of the assertions put forward was not always factually accurate. For example, one of the most widely circulated British pamphlets contained this sentence: 'From the assassination of the Archduke's wife at Sarajevo to the shooting of a hospital nurse in Brussels, Prussia and her Allies have concentrated their cruelty upon women.' Whatever his nationality, Gavrilo Printsip was certainly no Prussian, nor could he be said to have had sympathies with Austro-Hungary's cause.

British and Allied propaganda did not falter and stop, or worse still, retrace its steps to explain and defend. It moved forward and on, pouring out vast amounts of material and sticking to ruthless offensive tactics. The Germans were no less reluctant to use atrocity propaganda than the Allies, and their charges were no less extravagant and frequently as unsubstantiated by 'facts' which would stand close examination. But in their important campaigns they were not always on target, and if they were, their timing was at fault. Moreover, reports on atrocities were published by German government departments; hence they carried the stamps, not to say the stigma, of officialdom instead of enjoying the veneer of the prestige conferred on them by being written by independent scholars like Lord Bryce – however influenced such scholars may have been by the highly charged emotional climate of the time. Being official Government documents – and especially German documents – they tended to be enormously bulky and unwieldy and virtually unreadable. They were quite unfitted for popular distribution and consumption, unlike the short, neat pamphlets put out by the Allies. And, finally – most astonishing of all – most German atrocity propaganda material was not translated into foreign languages. The neutral world – one of the main target areas for this kind of propaganda – was blithely assumed to be able to speak and read German.

In short, the Germans did not lack the will to conduct this kind of propaganda; they lacked the capacity to translate this will into effective action.

CHAPTER 4 Courting the
United States

When the First World War broke out in Europe, it took most of the American public by surprise. There were some of course who favoured one side or the other, but the overwhelming majority wanted their country to remain neutral. It was part of American tradition to avoid getting involved in European wars, and this one seemed particularly complex in its beginnings and in the issues involved. It was best to stay well clear of any European entanglements.

This mood was reflected in President Wilson's first neutrality proclamation issued on 4 August 1914 and further underlined two weeks later when he appealed to the American people 'to be neutral in fact as well as in name'. In Paris a prominent American diplomat told the French historian Gabriel Hanotaux that 'there are perhaps 50,000 persons in the United States who feel that the nations should immediately intervene in the War on your side but there are over 100 million Americans who do not think so'. He added that it is 'our duty to reverse these figures so that the 50,000 may become 100 million'. But this last comment was his personal wish, the first was his objective assessment of the true situation of American attitudes which if one wanted to face realities, one dared not ignore. The diplomat's assessment was borne out by other evidence. Theodore Roosevelt, the former President, gave it as his considered judgement that the United States had best remain neutral. It was a judgement he was to alter subsequently when he came out in favour of the Allies, but it was the position. he held at the outbreak of the War.

Equally revealing of the state of American public opinion at the beginning of the War were the results of a questionnaire which a prominent American weekly magazine, *The Literary Digest*, sent to 367 American newspaper proprietors. The question asked was: Which side in the European struggle has your sympathies? The results showed that 105 editors favoured the Allied side, 20 the Central

Powers while 242, or almost exactly two-thirds of the total, expressed no particular preference.

In the light of the knowledge that the United States entered the War in April 1917, it may be difficult to imagine that in August 1914 America was overwhelmingly neutral and was determined to stay so. There was in 1914 no sign of an inevitable progression from neutrality to intervention. Many factors played a part in causing America to enter the War in April 1917. The propaganda of the belligerent powers was one of those factors, yet how decisive a factor must remain a matter of argument.

To help resolve the argument – without necessarily producing a satisfactory answer – it is essential to look at the propaganda efforts of the belligerent powers to win the favour of American public opinion and, if at all possible, to secure United States' active intervention on one side or the other.

Two of the belligerent powers effectively eliminated themselves from the struggle for American public opinion right from the start of the War and played no significant part thereafter. One was Czarist Russia, and the other was the Habsburg empire of Austro-Hungary. The first was an important, not to say, essential member of the Allies' war effort because it diverted a significant part of Germany's war potential from pressing home Imperial Germany's advantage in their attack through Belgium on France and Britain, and the second was not insignificant in continuing to tie up Czarist Russia's vast military forces in Eastern Europe when those forces could – from the Western point of view – have been more effectively employed to tilt the manpower balance on the Western Front.

From the propaganda point of view, neither Czarist Russia nor the Habsburg empire – the first allied to Britain and France, and the second to Germany – had popular propaganda appeal. To most Americans they represented all that was worst about the Old World, about Europe. Their rule depended on dynastic claims, backed up by rigid bureaucracies and ruthless, cruel police forces. Minorities were ignored, and if they protested at their treatment, suppressed.

It is, therefore, hardly surprising that, as far as their alliance with Czarist Russia was concerned, Britain and France tended to treat St. Petersburg as the capital of a country which, for some reason they were not quite able or wanted to explain, happened to be fighting on their side. By the same token, the Germans treated the Habsburg empire as a state which they had to support as loyal and honourable allies once the Habsburgs had got themselves into trouble. At the same time such support – which had to be given in honour-bound – did not imply that

Imperial Germany necessarily endorsed the Habsburg empire's policy of dominating all sorts of strange non-German, Slavic as well as non-Christian communities under its control. Austrians and Hungarians might have been exploited to win sympathy in the United States for the Central Powers, but Poles, Greeks, Slovaks, Croats, Slovenes, Serbs, Montenegrins and others who had settled in the United States, could hardly be expected to be any better disposed towards the Habsburgs than Jews or Ukrainians towards the Romanovs. The main belligerent powers on both sides of the First World War, therefore, tended to be silently discreet about two of their most important allies – in the Allies' case, Czarist Russia; in Germany's – Austro-Hungary.

The struggle for American public opinion thus rested primarily on Britain and France on the one hand, and on the other, on Germany. None of the three was happy about America's obvious determination in 1914 and early 1915 not to become involved in Europe's War. After all, the German, French and – after some heart-searching – the British public had become convinced that their cause was just and right, and that the enemy's was not only unjust but a menace to the fabric of civilized society on this earth. A cause so obviously right, each side argued, was bound to appeal to the people of the United States and the noble ideals on which their country was founded. If it did not, then Americans must have been misinformed or failed to understand what War was about. It was therefore the duty – backed, needless to say, by patent self-interest – of the main belligerent powers to battle for America's favour and, if possible, active support. Given the state of American opinion in 1914 – from the White House down to the tiniest hamlet – the two sides in the War started on more level terms in their struggle for American public opinion than America's entry into the War more than two-and-a-half years later in April 1917 might lead one in retrospect to suppose.

In theory, at least, the Germans should have started off with quite a few advantages. The most important was that of America's geographical position. While Britain hoped to be supplied with food and war material by the United States – as indeed she eventually was – and to be joined by America in the War against the Kaiser – as she was in 1917, – Germany could and should have concentrated on the less ambitious aim of keeping America neutral. This appeared a not by any means unattainable aim, given the state of American public opinion in 1914, yet Germany muffed it. Watching the performance of German propagandists in the United States, Count Bernstorff, the German ambassador in Washington until 1917, must frequently have been in acute despair.

Plainly he had little, if any, control over the personnel in charge of German propaganda in the U.S. The man who organized the German Press Bureau and Information Service in New York was Dr. Dernburg. He had been State Secretary at the German Colonial Ministry and had been sent to the United States to float a German loan. A determinedly neutral American Government warned against lending money to either side, and so Dr. Dernburg's plans came to nothing. He then revealed himself to be an agent for the German Red Cross and began collecting funds for that organization. At the same time he took it upon himself to explain the German case to America, and there is still some argument whether this activity was not the main purpose of someone of his seniority being dispatched to New York. Certainly he soon set up an organization with the help of the Hamburg-Amerika Line, German diplomats and businessmen stranded in the United States after Japan declared war on Germany on the 23 August, and unable to return to Germany, and a number of American journalists and businessmen. Daily news bulletins were issued as well as war pictures and film propaganda including film of German soldiers feeding Belgian and French children with captions such as 'Barbs feeding the hungry', or 'Dr. Dernburg's Barbarians look like this?'

In the distribution of its bulletins, pamphlets, pictures and films the organization set up by Dr. Dernburg could rely on the support of the German-American Alliance, which was well established in towns like St. Louis, Chicago, Cincinnati and Milwaukee. Furthermore, there was the Lutheran Church, with over 6,000 congregations in the United States whose parishioners numbered some three million. The services in many of these churches were still conducted in German, and a large proportion of their priests had been trained and ordained in Germany.

On the face of it, then, Dr. Dernburg's organization, given the limited objective of keeping America neutral – which it showed every indication of wishing to continue to be anyway, should have had a comparatively easy task, especially since it had readily accessible means of disseminating its message through the German-American Alliance and the Lutheran Church as well as the *New York Mail,* which was bought with the express purpose of reaching a metropolitan audience.

Why, in these apparently favourable circumstances, did the operation misfire? Broadly because it lacked subtlety, and particularly because it contravened every one of the four basic principles. In the first place, the message or messages which the German Information Service tried to get across, appeared disruptive. Because the objective Dr. Dernburg and his successors set themselves or were set, was limited, i.e. to keep America neutral, they imagined that their purpose might be best served by setting one discontented section of society against another – Protestant against Catholic, worker against employer, Jew against Gentile, Negro against White.

Their chief purpose in every instance was not to disrupt American society, but that was the impression they created. For example, the main aim in supporting the American Truth Society was to stir up the Irish Catholics against the British, and in addressing themselves to American Jewry they hoped to play on the fervent hatred of the Jews against the Czarist regime. 'It is impossible', the German Information Service announced, 'to be a comrade of Nicholas and not be a hooligan. In the days of Beaconsfield when England was far from Russia, no massacres of Jews were made, not on the poor, not on the rich. Today, when England is an ally of Nicholas, she must do as Nicholas does, she must make massacres, she must preach against the Jews.'

Dr. Dernburg's organization tried to play on the prejudices of every foreign language group in America and on real or alleged grievances of every minority group. None of these campaigns assumed a more absurd proportion than the German Information Service's drive to present Imperial Germany as the champion of the negro against his white oppressor, while in Europe Germany was calling the use of coloured soldiers by the French and British an 'atrocity'. Most extensive files were kept on the ill-treatment of negroes in the American South, and details of every lynching were sent to white and black newspapers in the United States. One can only guess what the editors of American newspapers made of the fact that they were being fed such information from such a source, but the over-all effect was that many Americans came to regard the activities of the German Information Service as 'foreign interference', un-American, unpatriotic and subversive. And since many brewers of German origin gave financial support to the German Information Service, it is perhaps not surprising that the investigations by the U.S. Senate in 1918 into German propaganda were conducted under the all-embracing heading of 'Brewing and Liquor interests and German Bolshevik Propaganda'.

German propaganda in the U.S. was equally neglectful of the second basic principle in the conduct of effective propaganda – analysis of the target area. The German Information Service repeated over and over again that the Belgian *franc-tireurs* were committing unspeakable atrocities against German soldiers and that civilian sniping was contrary to international law. Dr. Dernburg and his colleagues overlooked the fact that America is essentially a civilian country. If they had looked up American history books, they might well have concluded that the embattled farmers at Lexington, in international law, were civilian snipers, yet in American history books they are described as heroes. The Belgian snipers were regarded as heroes, too, despite what Dr. Dernburg said about international law.

Dr. Dernburg was as much at fault in his analysis of the German-Americans, of whom according to German calculations there were about 20 million in the United States at that time. He and his colleagues based their propaganda to them on the thesis that 'blood calls to blood'. As a result they were greatly disappointed. Of the 20 million German Americans not even a tiny fraction of them were willing to rally around *Kaiser und Vaterland,* when called upon to do so. What the German Information Service appeared to be unaware of, was that thousands of Germans had left for America after the abortive attempt in 1848 to build a united Germany on liberal principles. Others had left to get away from Germany's growing militarism, from its many kings and princes and its elaborate bureaucracy, others still to find better economic prospects. None had uprooted themselves, gone through the agony of adapting themselves to the climate of life in the New World and become Americans to turn their backs now on America at a distant Kaiser's command.

This kind of appeal, more than anything else, destroyed the image of credibility of Dr. Dernburg and his colleagues, the third requirement of effective propaganda. Such an appeal could not come from a friendly nation. It was meant to disrupt, to destroy.

To complete the picture, the Germans also flouted the fourth requirement of effective propaganda – communication. George Sylvester Viereck named his pro-German magazine *The Fatherland.* Where the British and the French asked for support in the name of humanity, democracy and common decency, Mr. Viereck asked for support in the name of the Kaiser. When Americans began to argue among themselves about whether their country should intervene in the interests of mankind and Americanism, Mr. Viereck argued about the interests of the Fatherland. The Germans did not learn that it would be far more effective to let an American put forward their case than to do it themselves. The New York office of the German Information Service was so busy writing its own propaganda that it failed to notice the favourable pro-German impact a wire sent to the Associated Press from Europe from five respected American journalists, including Harry Hansen and Irvin S. Cobb – all of them personally pro-Ally – had on American public opinion. The wire read:

'In spirit fairness we unite in declaring German atrocities groundless as far as were able to observe. After spending two weeks with

German army accompanying troops upward hundred miles we unable report single instance unprovoked reprisal. Also unable confirm rumours prisoners or non-combatants ... numerous investigated rumors proved groundless ... Discipline German soldiers excellent as observed. No drunkenness. To truth these statements we pledge professional personal word.' ˉ

Instead of giving this wire by Americans the widest publicity and repeating it over and over again, Dr. Dernburg and his colleagues were somehow convinced that their own speeches about the crimes of the *franc-tireurs* were more effective. Despite warnings Dr. Dernburg and his colleagues ostentatiously toured the length and breadth of the United States. Inevitably they were bound to overstep the limits of American tolerance.

It happened after the sinking of the *Lusitania* by a U-boat without prior warning, on 7 May 1915, with the loss of 1,198 lives, 128 of them American. Five days later the Bryce Report on Alleged German Atrocities in Belgium, was published. Wellington House had seen to it that large numbers of copies had been printed to ensure wide circulation in America. As stated in the previous chapter, the Report was compiled at an emotionally highly charged time, and incidents appended to it would probably not stand up to close scrutiny in a court of law in peace-time, but coming after the sinking of the *Lusitania* stories like this had a tremendous impact on American public opinion:

'Immediately after the men had been killed, I saw the Germans going into the houses in the Place and bringing out the women and girls. About twenty were brought out. They were marched close to the corpses. Each of them was held by the arms. They tried to get away. They were made to lie on tables which had been brought into the square. About fifteen of them were then violated. Each of them was violated by about twelve soldiers. While this was going on, about seventy Germans were standing round the women including five officers (young). The officers started it. There were some of the Germans between me and the women, but I could see everything perfectly. The ravishing went on for about one and a half hours. I watched the whole time. Many of the women fainted and showed no sign of life. The Red Cross took them away to the hospital.'

This was the moment that Dr. Dernburg chose to make a public speech in Cleveland, justifying the sinking of the *Lusitania* on the grounds that she was carrying weapons. Indignation in America ran so high that Dr. Dernberg was asked to leave the country.

The German ambassador, Count Bernstorff, probably felt no regret at seeing him depart and he insisted that Germans should in future be less prominent in American public life. In this he was only partially successful. And he had no success whatever in restraining the sabotage activities of an assistant military attaché in his embassy, Captain Franz von Papen who with his associates Bry-Ed and von Rintelen was carrying out a systematic campaign of trying to blow up American ships which were about to leave American ports with war material for Britain on board. So amateurishly clumsy were these activities that they were soon discovered, and it was suggested that the main motive of Papen and his associates was not sabotage, but to put the blame on the German-Americans and so to discredit their loyalty in the eyes of their fellow Americans and drive them back into the arms of the Fatherland. Whatever the motive – and Papen succeeded neither in his sabotage operations nor in discrediting the German-Americans – Imperial Germany's standing in American public opinion certainly sank by several degrees further.

Having started in August 1914 as one of the belligerent powers in a European war about which most of America did not want to know anything, and in which it certainly did not want to be involved, Germany was rapidly becoming the 'bad nation out', largely as the result of the actions of her own agents and representatives, and many Americans were beginning to call the Germans not 'Heinis' but 'Huns'.

The clumsiness of German propaganda may – so some cynics claim – well have pushed the United States into the First World War on the side of the Allies without any effort of persuasion in this direction on anybody else's part. Certainly German propaganda methods in seeking America's support or at least in securing her neutrality were singularly inept.

By contrast, Britain's efforts – whether by design or good fortune – were effective. Of course Britain enjoyed many advantages in her fight for American support which Germany did not. The two countries shared a common language, and Shakespeare and Dickens – not to mention living writers like H. G. Wells, Bernard Shaw, Arthur Conan Doyle and John Buchan – were better known to most Americans than Goethe, Schiller, Nietzsche, or Thomas Mann. But Britain's propagandists were not so foolish as to exploit this advantage by appealing – as Dr. Dernburg and Mr. Viereck did – to the superiority of their nation's culture or asking kindred blood to respond to the call of blood.

Unlike the German Information Service who in their appeals to the German-Americans happened to overlook the fact that the majority of the people to whom they were appealing had left the Fatherland because they did not like the conditions of life there, Britain's propagandists were acutely conscious of the events of 1776, when the inhabitants of the original 13 American colonies – mostly of Anglo-Saxon descent – took up arms to win independence from Britain. A call for support based on kinship was therefore likely to prove counter-productive.

A common language, a common culture may have been – and was – a tremendous advantage, but British propagandists were actively aware of the fact that if handled in the wrong way this advantage could easily turn into a disadvantage. It was therefore, in their view, much better not to ignore the American Revolution – which was a revolution against British rule – and the emotional attitudes which it had created in the minds of most Americans.

One of Britain's ablest propagandists, Ian Hay, tackled this dilemma when he wrote in *Getting Together* (published in New York in 1917) that America was more pro-Ally than pro-British because it believed that the Allies were on the side of right and justice in the War, but did not want self-satisfied John Bull to collect yet another scalp. He then went on to say:

> 'In this regard it may be noted that American school history books are accustomed to paint the England of 1776 in unnecessarily lurid colours. The young Republic is depicted emerging, after a heroic struggle, from the clutches of a tyranny such as that wielded by the nobility of France in the pre-Revolution days. In sober fact, the secession of the American Colonies was brought about by a series of colossal blunders and impositions on the part of the most muddle-headed ministry that ever mismanaged the affairs of Great Britain – which is saying a good deal. . . . In any case the fact remains that while in an American school-book the war of 1776 is given first place, correctly enough, as marking the establishment of American nationality, it figures in the English school-books, with equal correctness, as a single regrettable incident in England's long and variegated Colonial history. It is well to bear these two points in mind.'

The American reader is left to make up his own mind. There is no attempt at dragooning him into agreement, at a strident appeal for kin to support kin with the implication that failure to respond to the appeal amounts virtually to treason. Yet Hay's argument is couched in the highest tradition of Anglo-Saxon democracy, a tradition which both America and Britain share. On the other hand, it is not phrased in terms which underline or refer even vaguely to that common tradition.

On top of the many advantages which a common language, a common culture and shared traditions confirmed on Britain's propagandists, there was the hard, practical advantage that Britain controlled the transatlantic cables, the quickest and most efficient means of communications to North America. Germany eventually evolved a wireless service, but it was never as efficient and reliable as the cable service. As a result Britain throughout the War had the inestimably valuable advantage of getting her version of any event in first and so reducing Germany to denying or seeking to amend what Britain had said earlier. By this accident of news communication Britain was virtually assured of front page coverage of what she put over the cables while Germany was fortunate if her version rated one or two paragraphs, if any, several days later.

Yet the advantages which fortune gave Britain in the struggle for American support would have been of little avail if the people charged by the British Government with conducting this struggle and the way in which they conducted it, had failed to follow the ground-rules of effective propaganda.

The man whom Mr. Masterman put in charge of the division dealing with American propaganda at Wellington House, was Sir Gilbert Parker. Born in Canada, Parker had travelled widely in the United States, where he had a large circle of friends and was well-known through his writings. He had eventually settled in England, become a Member of Parliament in 1900 and had been knighted in 1902. He was known to have a deep and abiding interest in promoting better understanding between nations and once described 'the promotion of international goodwill' as one of the main aims in his life. His friends knew him as a kindly man who could usually be relied on to help in any worthy cause – he served on the executive committee of the Congo Reform Association – and in his judgement of his fellow men he preferred to take a generous and complimentary view rather than to couch his assessment in critical and derogatory words. When the *New York Times* asked him in 1913 – among other prominent figures in Britain and America – whether he would contribute an article to a series to mark the Kaiser's twenty-fifth anniversary of his accession to the German throne, he agreed immediately. In the article which appeared on 8 June 1913 he wrote: 'The highest praise I can offer concerning the Emperor William II is that he would have made as good a king of England as our history has provided, and as good a President of the United States as any since George Washington.'

Such was the personality of the man whom Mr. Masterman put in charge of supervising British propaganda to the United States. It is probably revealing of the way in which he approached his work that he called it not 'American propaganda' but 'American publicity'. In the British *Who's Who* of 1919 he had included one brief line about his war-time activity. It read: 'Had American publicity in charge for over 2½ years after war was declared.'

'Publicity' as against 'propaganda' may seem like a play on words, but it is more than that. It reflects an entirely different attitude of mind in tackling the problem of courting American support. Where Dr. Dernburg and his colleagues were intent on bludgeoning a rather confused neutral America into supporting Germany and the Central Powers and were intent on doing the job themselves without realizing that their rather too publicized efforts generated resentment among many Americans, Sir Gilbert Parker's policy was to choose the means which to him seemed the most effective in presenting the British case and then to leave it to the Americans to make up their own minds of where they stood in the War.

Consequently in choosing 'publicity' rather than 'propaganda' in defining his activities, Sir Gilbert Parker had to decide how best to put Britain's case across and to focus on the targets for his 'publicity' in the United States.

The key to the solution of the problem facing Sir Gilbert Parker is contained in a letter. Sir Edward Grey, then British Foreign Secretary, wrote to Theodore Roosevelt, dated 10 September 1914:

'My dear Roosevelt,
    J. M. Barrie and A. E. W. Mason, some of whose books you have no doubt read, are going to the U.S. Their object is, as I understand, not to make speeches or give lectures, but to meet people, particularly those connected with universities, and explain the British case as regards this war and our view of the issues involved.'

J. M. Barrie and A. E. W. Mason were not going to beat the British drum. How Theodore Roosevelt received them on arrival in the U.S. and to whom he chose to introduce them during their stay, was left entirely to the former President's discretion and mood. Needless to say, Sir Edward Grey may not have been entirely unaware of the fact that Theodore Roosevelt's attitude towards the War – a war in which only a month earlier, in August 1914, he had said the United States must remain neutral – was changing. Theodore Roosevelt had before the

War made no secret of his admiration of the German and the Kaiser and had proudly included in his Autobiography, published in 1913, the *Punch* cartoon of 16 November 1904, which pictured the Kaiser and President Roosevelt as 'Kindred Spirits of the Strenuous Life'. After Germany's sweeping military successes in Western Europe, he was, however, known to have begun brooding over America's future in case of total victory for Germany in Continental Europe. This, in Sir Edward Grey's view, was not the moment to attempt to push the former American President in the direction desired by the Allies, but at most to make him brood·a little more deeply. Here was a man who would support the Allies' cause, not in order to pull their chestnuts out of the fire but to preserve the America he cherished.

Sir Gilbert Parker took the same view as Sir Edward Grey. The purpose of British propaganda – or 'publicity' as Sir Gilbert Parker preferred to put it – was not to present Britain's case in terms of the cause of civilization. Whether Americans saw it that way, was entirely a matter for them. It was not for Britons or any other kind of non-Americans even to appear to tell Americans how to conduct their affairs.

Even as late as 1916 Sir Gilbert Parker encouraged Professor Gilbert Murray, the great classical scholar, to give expression to this point of view. At that time Professor Gilbert Murray wrote:

> 'It was not reasonable to expect the United States to plunge into war for motives of philanthropy. And if one begins to put the question on other grounds, then clearly it is not for us foreigners to decide what course best suits the interests or dignity of the United States. They know their own case, pro and con, far better than we can, and we certainly need not complain of either the skill or the fervour with which our friends in that great, strange country have stated our case.
>
> But the matter is decided. America will not join in this war. Both political parties are united on that point; and only a few voices of independent thinkers, voices sometimes of great weight and eloquence, are lifted in protest. I do not, of course, say that there might not arise some new and unexpected issue which would compel her to change her policy; but, as far as the issues are now known, the Americans have made up their minds to have no war.'

Wellington House recognized right from the beginning that the question of whether America should enter the War was a matter for Americans to decide, and that the best policy for non-Americans to adopt was a low profile of non-interference. Let Americans do the arguing, and concentrate on what Gilbert Murray called those 'of great weight and eloquence', to be supplied with the right kind of material to sustain their arguments.

In line with this thinking, Sir Gilbert Parker and his two chief assistants, Professor William Macneile Dixon, of Glasgow University, and Arnold Toynbee, the historian, began in September 1914 compiling a list of all the key political, academic, industrial, social, newspaper, financial and civic personalities in the United States. Their circle of personal friends and acquaintances was vast, and they supplemented that list by carefully going through *Who's Who in America*. By December 1914 they had compiled a mailing list which was in due course to include almost 250,000 people. The persons on that mailing list were sent thousands of pamphlets, cartoons, articles, speeches and photographs and, where appropriate, films. Every one of these communications was accompanied by a card stating that it had been sent 'with the compliments of Sir Gilbert Parker' and with an invitation to let Sir Gilbert know how he or she felt about the material he had received.

The whole elaborate operation was kept on a personal basis. Wellington House was not ever mentioned. Indeed, this practice was broken only after the United States entered the War in 1917, and Professor Dixon had taken over from Sir Gilbert Parker and used the Wellington House address, 8 Buckingham Gate, on his enclosure cards.

Sir Gilbert, with his not inconsiderable experience of the United States, also showed an acute awareness of the importance of small town and country newspapers. Hundreds of pamphlets and leaflets were dispatched to them every week 'with his personal compliments'. Needless to say, the smaller newspapers appreciated this kind of attention even more than the larger ones. They, therefore, frequently took more trouble to reply to Sir Gilbert Parker than the big city publications and as a result Sir Gilbert could supply the British Cabinet every week with a fairly shrewd analysis of some depth of the temperature of American opinion.

He could also supply the American press – both the mass circulation papers and the local publications – with articles by writers like G. K. Chesterton, Sir Arthur Conan Doyle, Alfred Noyes, Joseph Conrad, H. G. Wells, Bernard Shaw and Rudyard Kipling. Not only were they good writers, but Sir Gilbert Parker did not make the mistake of trying to censor what they wrote. What mattered to him was not that H. G. Wells was critical of some aspects of life in Britain; what mattered to him, was that no matter how critical Wells might be of Britain, he made it very clear that he would rather have Britain win than Germany

Hindenburg: „Majestät, das Volk ist gedrückt und murrt unaufhörlich."

Majestät: „Weshalb murren sie? Wir spüren keine Last."

9. Hindenburg warns the Kaiser that the German people are groaning under
   their burdens. The Kaiser replies: 'Why are they groaning? WE feel no
   hardship'.

10. Exaggerated anti-Germany propaganda published after the First World War antagonized many Britons and so inadvertently helped Hitler.

because a German victory would be a victory for militarism and under militarism society was unlikely to be improved.

With writers of such quality, Sir Gilbert Parker had little difficulty in placing their articles. Even the Hearst newspapers, which were frequently suspected of pro-German sympathies, bought scores of their articles.

Sir Gilbert Parker adopted a not dissimilar attitude towards the Union of Democratic Control in Britain, which was run by men like H. N. Brailsford, and towards British Labour and trade union leaders, as he adopted towards H. G. Wells and Bernard Shaw. Very well, Britain and France might not be perfect but would victorious Imperial Germany give them a chance to improve matters? Why did they not travel across the Atlantic and discuss the point with their colleagues in the American Federation of Labour? They did, and there is evidence that quite a few American labour and trade union leaders were strengthened as a result in their belief that an Allied victory was preferable to a German one.

It was only among America's Jews that Wellington House made little progress. Not that it ever tried very hard. It knew that the hearts of those Jews who had any relatives in Czarist Russia could be expected to warm to the cause of an alliance which included the traditional land of the anti-Jewish pogroms. It also knew that most liberal-minded people – whether in America, Britain, France or elsewhere – would not be willing to swallow the picture of Czarist Russia as a champion of humanity and civilization. By contrast, Dr. Dernburg and the German Information Service won some friends among American Jews because Germany before and during the First World War had treated the Jews well. Germany, in fact, was the land which offered refuge and rest to many Jews who managed to get out of Czarist Russia and remembered it with gratitude as a comparatively friendly place. The German Information Service, however, failed to follow up its advantage. It was too concerned with projecting German *Kultur,* denying Allied accusations of German bad conduct and seeking to set America's many minority races against each other. In any case, such misgivings as some of America's Jews may have felt about the Allied cause were largely dispelled when, in November 1917, A. J. Balfour, then British Foreign Secretary, committed Britain to the establishment of a Jewish National Home in Palestine. America was by then in the War, Russia had become a completely unreliable ally, and the most important impact the Balfour Declaration had was on the Jews of Central Europe at a stage when war weariness was becoming increasingly noticeable. General Ludendorff described the Balfour Declaration as the cleverest propaganda stroke of the Allies and asked disconsolately why Germany had not thought of it first. What he did not ask or answer was whether, if Germany had thought of it first, it would have made any difference.

Wellington House may have done its best to play down the alliance with Czarist Russia. It did not try to avoid the issue of Britain's naval blockade of Germany. The blockade was a sore point with many Americans. It hit many of them – producers of raw materials, manufacturers, brokers and shippers – in their pockets. Of course in the long run it tied America's economy increasingly to that of Britain and her Allies. But it was an emotionally explosive issue, especially as it formed the justification for Germany's use of unrestricted U-boat warfare. It therefore had to be answered. It could not be ignored.

On this issue Wellington House used the same method of pamphlets, letters and articles sent personally to thousands of people of 'weight and influence' up and down the width and breadth of the United States. It was pointed out very delicately that during the American Civil War many people in Britain had been for the North, that the North had blockaded the South and that a little reciprocity at this time might not be entirely unfitting. Mr. Masterman was one of the many propagandists at Wellington House who put forward Britain's views in the matter. In 1915 he wrote:

> 'Sooner or later the Central Powers will be in the position of the South in America.
>
> I believe every unprejudiced person will agree that in blockading Germany, in preventing neutrals from violating the law of blockade and providing Germany with the food and other articles she needs to carry on the war, Great Britain instead of having exceeded her legal rights has surrendered many of them, and that for everything she has done she is following the precedents established by the United States during the Civil War, which the Supreme Court of the United States sustained as being in strict conformity with good morals and international law.'

Wellington House not only followed the method of using persons as channels of influence by providing them with arguments, couched in terms of objectivity and reason, to decide whether or not America should intervene in the War for her own sake and for the sake of humanity. It also appealed to a wider public in terms which made no attempt at objectivity but were aimed to portray the enemy of the Allies as a brutal barbarian whose defeat was essential to the survival of civilized life on this planet.

The themes are clear in retrospect, and their effect on public opinion was cumulative: there was the rape of brave little Belgium and the German atrocities as detailed in the Report of Lord Bryce, a former British ambassador to Washington and the distinguished author of 'The American Commonwealth'; there was the sinking of the *Lusitania* which after repeated protests by President Wilson eventually led the German Government to suspend – at least temporarily – unrestricted U-boat warfare; there was the execution of Nurse Edith Cavell; there was the announcement by the German Government that on and after 1 February 1917 unrestricted U-boat warfare would be resumed.

On 3 February President Wilson broke off diplomatic relations with Germany and handed Count Bernstorff, the German ambassador, his passports. Sir Gilbert Parker, who happened to be in the United States at the time, resigned from his post partly because he was in bad health but largely because he felt that his work was done.

Yet there was still a short way to go before the United States entered the War. The anti-interventionists were fighting a stubborn rear-guard action. The moment of decision did not come until British Intelligence intercepted a message from the Imperial German Foreign Minister, Alfred Zimmermann, to Mexico, hinting that should a rupture take place between Germany and the United States, an alliance would be desirable between Mexico and Germany. This intercepted message was immediately passed on to Washington, which released it to the American people on 1 March. Public indignation reached boiling point. The prophecies of the most ardent interventionists that Germany intended to turn America into a German colony seemed confirmed.

On 2 April President Wilson asked Congress to declare war. The vote for war in the House of Representatives was 373 to 50 and in the Senate 82 to 6. On 6 April President Wilson issued a formal declaration of war.

It is none the less impossible to calculate exactly how much Britain's propaganda efforts contributed to this decision – or, for that matter, what part the clumsy ineptitude of German propaganda played.

CHAPTER 5

Undermining Enemy
Morale

In total wars, such as the First and Second World Wars, in which
sooner or later every section of the society of the belligerent powers
becomes involved, it is inevitable that one's own side is, in the tradition
of Western films, depicted as 'goodies' and the enemy side as the
'baddies'. Each side feels compelled to prove that its own cause is just,
that the enemy's is dastardly, barbaric and inhuman, and that victory
for one's own side is certain.

In the First World War the propaganda efforts of the Allies and of
the Central Powers varied in the emphasis contained in the 'message'
which was put across in these three themes. The emphasis depended to
no small degree on the military situation prevailing at the time. As a
rule, the better the military situation seemed to be, the nobler were the
moral principles one side or the other appeared to be fighting for and
the stronger the appeal to the other side – at least important sections of
it – to give up the unequal struggle. For example, on 16 August 1914
when Russian troops stood deep in East Prussia, the Czarist Government announced that it proposed to grant autonomy to a united
Poland. There were of course, a great many Poles fighting in the ranks
of the Central Powers on the Eastern Front. Nicholas II, the Czar of all
the Russias, in part, happened of all rulers to be the man to anticipate
some of the Fourteen Points which President Wilson was not to
enunciate for another three years – by which time the Czar had lost his
throne.

Again, Britain, France and Imperial Russia were extremely reluctant
to go back on their secret and not very creditable undertakings to Italy,
made to that country in 1915 to induce her to enter the War on the
Allied side, which would have meant that substantial parts of Croatia
and Dalmatia would have been handed to her despite the fact that these
territories were predominantly Slav. Only after the United States had
entered the War and President Wilson had publicly insisted on the
principle of self-determination, and only after Italy's armies had

suffered a nearly catastrophic defeat at Caporetto in 1917 at the hands of the Austro-Hungarian forces, did Italy agree to the creation of a Yugoslav state after the War and thus make it possible for the Allied propagandists to exploit the disaffection of the non-German and non-Magyar subjects of the Austro-Hungarian empire.

The military situation plays a crucial part in moulding the psychological climate in which propaganda aimed at undermining the enemy's morale is likely to be most effective. It is this interplay of forces which America's entry into the War in 1917 affected so decisively. In the military sense American entry tilted the balance in favour of the Allies provided American forces could be brought to Europe quickly enough; in the propaganda sense President Wilson could appeal to minorities within the enemy countries and even to Germans and Austrians over the heads of their Governments. America's entry brought about an irresistible combination of superior military force and the highest and most noble human sentiments.

Undermining enemy morale is never easy, and it was particularly difficult in these early years of the First World War. All the belligerent countries had entered the War on a crest of patriotic fervour. In Britain alone there had been a brief moment of hesitation, but once the doubts had been resolved by Germany's rape of 'brave little Belgium', Britain's determination to see it through was as strong as that of the other belligerents. In such an atmosphere even the most brilliant propagandist would have had a difficult time to undermine enemy morale because undermining enemy morale means shaking one's opponent's confidence in victory and his determination to fight on to win.

To shake Germany's confidence in victory in 1914 was rather like trying to push a pea up the slippery slope of a glass mountain with the tip of one's nose. Still, the British tried. In October 1914 their aeroplanes scattered what were probably the first leaflets to be dropped on German troops in France in the First World War. It told them that the Russians were actually occupying German soil and that the German army had sustained losses of 70,000 men. The concept behind the leaflet was sound: East Prussia had been invaded by the Russians; the German drive to seize Paris had been halted because, as most German soldiers suspected, troops urgently needed in the West to press home their advantage, had to be diverted to the East. The leaflet, therefore, touched on a sore point, but despite the soundness of the concept behind it, there is no evidence to suggest that it had much, if any, impact on German troops. In their view, the 'Tommies' were just trying to be clever. After all, had the 'Tommies' not heard of Hindenburg's and Ludendorff's brilliant victory over the Russians at Tannenberg? Moreover, although the German army's advance in the West may have come to a halt for the time being, German troops were still within striking distance of the French capital. Victory may have been delayed by a few weeks or months but it remained as certain as ever.

The leaflet illustrates the dangers of failing to study the state of mind of the target – in this case German troops in France. It also illustrates one essential requirement in all good propaganda – credibility. By ignoring Germany's victory over the Russians at Tannenberg, the leaflet lost all credibility in the eyes of ordinary German soldiers.

In the early days of the war the British also overemphasized and even exaggerated the size of the Anglo-French forces in the West. They sought to increase the number of men facing the Germans by propaganda. Their aim was to blur the number of British soldiers actually engaged in fighting in France with the number flocking to enrol and being trained in Britain and soon to be sent to France. It was a ruse with which those in charge of French propaganda, were in violent disagreement. The Germans, they argued, would not be fooled, and in the meantime the Allies were in retreat almost everywhere. If the German advance was not halted, was not this line of propaganda likely to boomerang? Moreover, was it not true that the Germans knew exactly how strong the British Expeditionary Force in France was? Had the Kaiser not referred to it as 'contemptible'? It is highly doubtful if the Kaiser ever made such a remark.

Though a man given to impulsively making remarks on anyone and anything without thinking of the consequences these may cause, he was known to have a high regard for the British army and the British soldier. He was extremely proud of the honorary colonelcies he held in a number of British regiments – until the outbreak of War, – had his British uniforms valeted with the same care as his numerous German uniforms and had, before August 1914, never failed to attend a ceremonial function of a British regiment with which he was associated.

If he made any reference to the British Expeditionary Force of 1914, it is unlikely he used the word 'contemptible'. He may in German have used an expression like 'ridiculously small' – which the B.E.F. was, after all, in comparison with the vast conscript armies of the Continental powers – and this expression may easily have been translated into English as 'contemptible'.

Whatever the truth, the Kaiser's alleged remark made tens of thousands of young Britishers rush to volunteer for the Army. It was more effective than many a recruiting slogan. And since the German advance was halted before Paris, it was German propaganda that had to switch from predictions of immediate victory to the theme of hopes deferred, while the misgivings of the French proved groundless, and reality was given a chance to catch up with propaganda. Indeed, as the War on the Western front dragged into stalemate, French and British propaganda joined together in asking the Germans over and over again: Can you ever win? Are you not doomed to fail?

As the War continued, the propagandists devised many different ways of undermining the enemy's morale by sowing doubts and discontent and so nibbling away at the will to fight on. Before the Christmas season of 1915 the French brought out a tract which they called *Die Feldpost*. It was dropped over large sections of the German lines, but there is no evidence to suggest that many Germans accepted it as a genuine German Army publication and failed to realize where it came from. Yet this did not affect the tract's credibility because it dealt in very simple human terms with the pleasures of spending Christmas at home in the circle of one's family. This was the second Christmas the Germans who, as the French knew, regarded the festive season with particular devotion, had to spend away from home, and victory seemed as far away as ever. Even though most Germans knew where the tract came from, they were mesmerized into reading over and over again what they knew they would be missing.

The tract did not cause mass desertions, but equally it did not raise the spirit of the German soldiers who read it.

Later on in the War, the French dropped propaganda sheets on the German lines called *Briefe aus Deutschland* which purported to be addressed by German families at home to menfolk on the front. Again, the Germans were not taken in by the alleged source of the *Briefe*, but again that did not affect their credibility. The *Briefe* dealt with the near starvation rations of the civilian population at home, the absence of essential goods in the shops, especially clothes, the appalling conditions of everyday life. The Germans knew that the food rations their families were receiving were low, and although they may have suspected that the *Briefe* may have made out the situation to be worse than it was, essentially what they read in them was true.

The French were the first among the Allies to set about driving a wedge between the Kaiser and German militarism on the one hand and on the other the German people. One of the French leaflets scattered over Germany showed on one side a picture of the Kaiser and his numerous strapping sons, resplendent in their uniforms, unscathed by war, and on the other side endless rows of wooden crosses marking the final resting places of his loyal subjects. Another French leaflet carried on one side a picture of the Kaiser and his generals, sitting at a table and talking amicably over a convivial drink; the picture on the other side showed a direct hit on a German trench with countless soldiers being ripped to pieces in the explosion.

On the instructions of George Creel, President Wilson and the Chairman of the Committee on Public Information, the theme of driving a wedge between the Kaiser and the Germans became one of the main strains in American propaganda once the United States entered the War in 1917, as it did in British propaganda once Lord Northcliffe was appointed Director of Propaganda in Enemy Countries in 1918.

Not that the Germans did not try their hand at trying to drive a wedge between governments and people. Their particular target for this kind of propaganda were the French rather than the British. Frenchmen, the Germans argued, are usually prepared to believe the worst of their rulers. If therefore German propaganda plugged away at corrupt politicians living it up in Paris and receiving enormous bribes from munition manufacturers and munition manufacturers making extra profits by supplying the French army with below standard weapons, they would surely be merely telling Frenchmen what many of them half-believed already. In fact, this line of propaganda had remarkably little effect – except perhaps in April 1917 when several army corps of the French mutinied after having sustained staggering losses in a completely abortive offensive launched by General Nivelle. But even then when the target, in this case the French soldier, was likely to be more receptive than at any previous time, the impact of this form of propaganda was not great. One of the reasons – and perhaps the most important one – was that the Germans throughout the War did not supply their propaganda services with adequate resources. The hundreds of thousands of leaflets which should have showered on demoralized French troops after the failure of the Nivelle offensive and before General Pétain could restore discipline, had not even been printed, and had they been, there were no practical means available to deliver them in the desired area.

One of the means which the Germans used to disseminate propaganda among the French, was the *Gazette des Ardennes,* a paper printed for the French population in German-occupied France. The Germans ensured that the French bought the paper by publishing in it lists. of French soldiers captured by them. No French family with menfolk serving in their country's army could be expected to resist such bait. And in the *Gazette* the Germans planted many stories which they knew would trickle through to the rest of France somehow. But in propaganda – as Lord Northcliffe pointed out – a trickle is not enough. And the *Gazette* was rather like blowing a tin whistle in a gale and expecting to be heard.

The Germans tried various other means of disrupting French morale. The *Gazette* published accounts of Moroccans who had had babies by the wives and sweethearts French soldiers had left behind. And they tried to revive the ancient feud between Britain and France. Britain, it was alleged, was fighting to the last Frenchman. British soldiers were, if German propaganda was to be believed, having even more babies by French soldiers' wives and sweethearts than the Moroccans. And Britain was determined to annex Calais after the War.

The Germans did not make any noticeable progress in their campaign of splitting the Allies on the Western Front. By contrast, Allied propaganda on splitting the Central Powers was more effective. It told the Germans in France that the Austrians and Hungarians hardly knew what food rationing meant, but had it ever occurred to them to send any food to Germany? Furthermore, the Allies said that Austrian and Hungarian soldiers had become experts at avoiding battles. Did German troops not have to be sent to Russia and even to Italy whenever the Austro-Hungarian armies were in trouble?

This campaign of disruption may not have undermined the German soldier's determination on the Western Front to go on fighting, but it certainly increased his grumbles about his ally. The campaign was effective because it was credible – at least one half of it. The Germans knew that German divisions had to be despatched constantly at short notice to stiffen the Austro-Hungarian forces on various fronts, and they had no means of checking on the food situation in the Austro-Hungarian empire. So the truth of one part of the Allied propaganda campaign rubbed off on the other part.

The campaign of splitting the Central Powers was directed not only at the Germans but also at the Austrians. Had the Prussians not always had nothing but contempt for the easy-going Austrians? Did the Austrians not know that Berlin had plans to take over large parts of the Austro-Hungarian empire? The Dual Monarchy was not Prussia's ally, it was her satellite.

Such propaganda fell on receptive Austrian ears. The German units which were sent to stiffen the Austro-Hungarian forces on different fronts did not always behave with the tact and consideration one would have expected from someone coming to the aid of an ally, and so many Austrian officers and men were inclined to believe what the Allies told them. Their morale, already low, sank lower still as a result of this Allied propaganda campaign.

Disruptive political propaganda in total war can be measured, if it can be measured at all in terms of effectiveness, in relation to the military advantages it gains. The first to win substantial advantage in this sphere were the Germans against Czarist Russia. They may have lost the advantage they gained in the long run, but their gain was substantial, at least to begin with, and if they failed to exploit it, they had only themselves to blame.

Certainly, their campaign showed great imagination. Faced with a war on two fronts – a situation Bismarck had always warned against – they soon discovered that the Czarist armies were poorly equipped, poorly clothed and poorly led. They could be out-gunned, out-manoeuvred and out-generalled by any disciplined and well-led body of men, irrespective of Russian superiority in manpower. The only problem was that this would take time. And Imperial Germany did not have the necessary time. It needed the German troops tied up on the Eastern Front in the West. For it was in the West that the issue was going to be decided.

Given this situation, Hindenburg and Ludendorff had to think of a way of bringing the war in the East to an end and thus transferring the German troops in the East – about two millions – to the West, to give Germany an overwhelming numerical advantage over the Western Allies and thus ensure the defeat of the Allies and Imperial Germany's victory.

The result was, to put it mildly, bizarre. Czarist incompetence stimulated the movement for change in Russia, but since the autocratic nature of Czarist rule made it impossible for any change to be other than revolutionary, German generals who had turned purple with rage when a Social Democratic member of the Reichstag had tabled a motion proposing some mild form of social reform, began actively to support social revolution inside Russia. The Junkers and the Reds – admittedly Russian Reds – became allies. In Russian prisoner-of-war camps the German placed revolutionary pamphlets, advocating the overthrow of Czarist Russia, in the hands of Russian prisoners. These pamphlets said nothing that was complimentary about the institution of the monarchy and the allegedly privileged position of big land-owners. They were aggressively scurrilous about big industrialists and their treatment of workers. A Russian Red Cross official who visited these camps under the auspices of the International Red Cross protested vigorously against what he regarded as the unscrupulous exploitation for propaganda purposes of his country's prisoners of war. His protests against these systematic attempts to poison his captured fellow-Russians against their country's lawful Government produced no results whatever. On the contrary, throughout 1916 the German generals stepped up their efforts to have Russian prisoners-of-war exchanged or released as quickly as possible so as to ensure that they would return to their homeland to spread the revolutionary word. The members of the German General Staff who laboured so assiduously in this field, were themselves largely members of the landowning aristocracy with large estates and frequently close family ties with big business.

In March 1917 the revolution which Germany had helped to bring about swept Nicholas II from the throne. Kerenski formed a provisional government and announced that Russia would continue the War against Germany at the side of her allies. Ludendorff, determined to extricate Germany's armies in the East and bring them West as quickly as possible to secure a decisive victory in France, had Lenin taken from Switzerland across Germany to Russia in a sealed train. But Lenin did not seize power in Russia until November 1917, and peace negotiations between Germany and Russia began at Brest-Litovsk in December. At last Ludendorff seemed free to reap the benefit of his unorthodox propaganda and psychological warfare initiatives. Another two million German soldiers in the West would give the Central Powers the numerical superiority they had lacked since 1914 – before America's power could make itself felt – and promised victory at long last.

It was not to be. The peace terms which Germany imposed on Russia at Brest-Litovsk, would have meant that Russia would have lost a third of her population, a third of her agricultural land and more than half her industry. To enforce these terms Ludendorff had to leave more than 800,000 German troops in the East. Only just over a million could be moved to the West. They gave Ludendorff a numerical superiority over the Allies but, with America's forces increasing week by week, not a large enough one. He launched his spring and summer offensives in 1918 and won much valuable ground but failed to attain his ultimate objectives. And then the Allies, reinforced by the Americans, counter-attacked. Ludendorff's cynically imaginative disruptive propaganda drive had failed in helping him to achieve his military ends as it was meant to.

Curiously, the Austrians were more successful than Ludendorff in their attempt at disruptive political propaganda in support of their military operations. Admittedly, their attempt which preceded the battle of Caporetto in October 1917 on the Italian front, was less grandiose in design, but it was more carefully prepared, planned and executed than Ludendorff's.

The Austrians made a meticulous analysis of the state of mind of their target, the average Italian conscript. With the exception of those who served in a few crack regiments, he was war-weary in the extreme. He had fought innumerable battles on the Isonzo without either side gaining any substantial advantage and with both suffering substantial losses. His rations were poor and did not always get to him. He felt disgruntled because his officers seemed better-clothed, fed and quartered. His leaves home were few and far between and liable to be cancelled at the last moment without his being given any adequate reason. Yet he suspected that his officers went off on leave practically whenever they felt like it.

The Austrians sought to take advantage of this state of mind by showering the Italian lines with leaflets which purported to be letters to the men from their families pleading with them to come home. There was not enough food at home, the leaflets claimed. The allowances the Army paid to families were inadequate, and inadequate as they were, there usually was some official who found a reason to make some deduction and to make them more inadequate still – probably to line his own pockets.

Whether believed or not in whole or in part, this was hardly the kind of stuff to put fire into the belly of a soldier before battle.

Moreover, the Austrians had a second string to the bow of their disruptive propaganda campaign before Caporetto. It was socialism. Horrified at the vaguest stirrings of even the mildest radicalism inside the Dual Monarchy, their military intelligence officers suddenly developed a sensitive nose for social injustice, for the enormous gaps between the rich and the poor, for the causes of discontent in Italy. More and more Italians, their intelligence analyses showed them, were turning towards Socialism and Communism in despair as offering the only way out of Italy's social inequalities. Here, surely, was a movement that called for encouragement. And so Austrian officers who would not tolerate anyone suspected of being a Socialist to walk the streets of Vienna freely, brought Austrian soldiers who had caught the 'Bolshevik bug' on the Eastern Front to Italy, and encouraged them to mingle, whenever opportunity offered, with Italian soldiers. Of course they ran the danger of being shot if caught by an Italian officer or non-commissioned officer, but the terrain was on their side, and to many the chance of converting a few Italian comrades to their way of thinking seemed preferable to lingering for years in an Austrian gaol. The Austrian authorities released even a few Austrian pacifists from prison and asked them to mingle with Italian soldiers. In short, no effort was spared to make the disaffected Italian conscript feel more dispirited.

No doubt these efforts at undermining Italian morale contributed to

Italy's disastrous defeat at Caporetto, when the Italians lost over 300,000 men and were driven back 70 miles to the Piave River. How much they contributed, is bound to be a moot point. Certainly General Cadorna, then the Italian Commander-in-Chief, had not disposed his armies in a way that was likely to ensure a successful defence.

The success the Austrians had in shaking Italian morale – whatever its precise contribution to Italy's defeat at Caporetto – was within a few months, however, to be overshadowed by what was probably the biggest single propaganda triumph of the War. The target was the Austro-Hungarian forces which had won the battle of Caporetto and had advanced to the banks of the Piave, where they were preparing for their next drive against the Italians. That drive was to be launched in April 1918, but so intensive and effective was the Allied propaganda offensive against the Austro-Hungarian forces that the Austrian command was compelled to postpone its attack until June.

The purpose of the Allied campaign was to divide the non-German and non-Magyar subjects of the Austro-Hungarian Empire from their Austrian and Hungarian Masters. Professor Masaryk, the president of the Czecho-Slovak National Council, had already received an assurance from President Wilson, backed by similar assurances from the other Allies, that an independent Czecho-Slovak state would be set up after the War. The Poles' claim to a separate state had also been recognized. The stumbling block to the launching of an effective propaganda campaign to win the loyalty of all the subject Habsburg nationalities was the Southern Slavs. Much of their homeland had been offered to Italy in a secret treaty concluded early in 1915 in London, in order to induce Italy to enter the War on the Allied side. The promises in this treaty were completely incompatible with the unification of all the Southern Slavs in one state, in one Yugoslavia, but the Italians insisted on the implementation of the promises made to them. They preferred annexation to the problematical friendship of what they regarded would become merely a much enlarged Serbia in which the other Southern Slavs would be treated as inferior citizens. The Southern Slav Unitary Declaration signed by Trumbitch, the president of the Southern Slav Committee, and Pashitch, the Prime Minister of Serbia, on Corfu in June 1917, put paid to the last mentioned Italian argument. The Declaration also contained a warning that all the territory compactly inhabited by the Southern Slavs 'cannot be mutilated without attaint to the vital interests of the community. Our nation demands nothing that belongs to others, but only what is its own'. At the same time Wickham Steed, who had for many years served as *The Times* correspondent in Vienna and was later to become editor of *The*

Times, and Dr. Seton-Watson pleaded the cause of the suppressed nationalities in the Habsburg Empire not merely for the sake of propaganda but in the interests of justice and self-determination. In America, which had been in the War since April 1917, George Creel, the head of the Committee on Public Information, who was responsible for propaganda both inside and outside the United States, pointed out that many who belonged to the suppressed nationalities within the Habsburg Empire had emigrated to America. There were among them numerous Southern Slavs and these could hardly be expected to welcome the prospect of tens of thousands of their relatives being cold-bloodedly handed from one foreign ruler to another. Nor would other Americans, even those with no ties with Central or South Eastern Europe, like to see their country involved in what they were bound to regard as a revolting piece of heartless European diplomacy. Creel, who fortunately had ready access to President Wilson, was able to put his case personally to the man in the White House, and President Wilson came out strongly for the principles of self-determination and of 'government by consent of the governed'.

Still the Italian Government held out. It was not until after Caporetto that it changed its mind, and agreement was reached with the Slavs. The last stumbling block to sow disaffection among the suppressed nationalities in the Habsburg Empire and win their support for the Allies had been removed.

The opportunity which was thus opened, was summarized by Lord Northcliffe, Director of Propaganda in Enemy Countries, in these terms:

'The Empire of Austria contains some 31,000,000 inhabitants. Of these less than one-third, i.e. the 9,000,000 or 10,000,000 Germans of Austria, are pro-German. The other two-thirds (including the Poles, Czecho-slovaks, Rumanians, Italians, and Southern Slavs) are actively or passively anti-German.

The Kingdom of Hungary, including the "autonomous" kingdom of Croatia-Slavonia has a population of approximately 21,000,000 of which one-half (Magyars, Jews, Saxons and Swabians) may be considered pro-German, and the rest (Slovaks, Rumanians and Southern Slavs) actively or passively anti-German.

There are thus in Austria-Hungary, as a whole, some 31,000,000 anti-Germans, and some 21,000,000 pro-Germans. The pro-German minority rules the anti-German majority.'

The proportions in the population as a whole were broadly reflected in the composition of the Austro-Hungarian army, and Wickham

Steed and Seton-Watson, whom Lord Northcliffe had put in charge of the Austro-Hungarian department at Crewe House, decided that a British-French-Italian-U.S. body, naturally also containing representatives from the suppressed Habsburg nationalities, must be established in Italy immediately to launch an intensive propaganda campaign against the Austro-Hungarian forces on the Piave front. Incidentally, one of the first suggestions which George Creel's representative made to Washington was that a few token contingents of American soldiers should be sent to Italy, because many Italians still did not believe America was already actively engaged in the War in Europe. The suggestion was not popular with the Allied command on the Western Front, who wanted every available soldier in the war area in which they felt the final decision would fall, but George Creel prevailed. American soldiers were sent to Italy, enough to be seen by the Italians but not enough to be able to have an important part in deciding the outcome of the ensuing battle of the Piave. Propaganda was to have a more decisive influence on that battle.

The Inter-Allied Commission, comprising representatives of the subject nationalities of the Habsburg Empire, acquired printing presses in Reggio Emilia, which produced a weekly bulletin in Czech, Polish, Serbo-Croat, Slovene and Rumanian. It also produced thousands of leaflets and coloured reproductions of a patriotic or religious character which appealed to the nationalist and religious aspirations of the suppressed nationalities. The message was simple: Why fight and perhaps die for a foreign oppression? Why not join the Allies and fight to make your homeland your own? Is it not time at long last, after centuries of oppression, that your native land should be yours, that you should have a say in its future?

All this vast amount of material was delivered straight from the printing presses to the Italian forces on the Piave. Every Italian army corps had an aeroplane attached to it which had no other duty but to scatter the leaflets over the Austro-Hungarian lines. Rockets were also used to distribute the leaflets and pamphlets, which soon ran into millions, and so were grenades. But the most effective distribution was by means of so-called 'contact' patrols. These patrols consisted of Czech, Polish, Croat and Serb dissenters from the Austro-Hungarian forces who had volunteered for the job of moving into 'No-man's land' and trying to establish contact with fellow countrymen. When they were successful in making contact, they naturally did not confine themselves to merely handing over the leaflets they were carrying; they discussed the future of their native land, their own future with their fellow countrymen. Sometimes a number of the men who had talked to a 'contact' group, joined that group on the spot and deserted, while the rest crept back to the Austro-Hungarian lines to spread the word.

Gramophones were also pressed into service. They were placed in 'No-man's land' and played records of nationalistic or nostalgically sentimental Czech, Slovak, Polish, Serb, Croat or Slovene songs, and the Italians frequently heard men on the other side join in the song or clap their hands to the tunes.

This propaganda onslaught and its effect caught the Austrian command unawares. The Austrian generals did not know from day to day how many men would still be there to answer the next morning roll call. Munitions dumps exploded mysteriously. One Slovene company commander, making his evening tour of inspection, overheard his men talking of going over to the other side. He told his men he shared their views and during the night led them across as a fully-equipped company. The Austrians in the morning found a gap in their front line.

The Austrians discovered that two-thirds of their forces had to be considered unreliable. They had no means of telling how many of what after all constituted a majority in their army, was disaffected, how many on the verge of desertion or mutiny. Plainly their forces were in no condition to launch the offensive on the Piave front in April as planned. The attack had to be postponed for at least two months. In the meantime, the Austrian command tried to re-organize its forces. Czech and Southern Slav units were 'mixed' with reliable Austrian and Hungarian units. In June machine gun posts, manned by either Austrians or Hungarians, were set up all along the front to prevent unreliable troops from deserting.

In the long run, it was all to no avail. The Austrian Piave offensive when it was eventually launched in the second half of June, ended in fiasco. The Battle of the Piave demonstrated what was soon to become obvious to the whole world: the Austro-Hungarian army had ceased to exist as a cohesive coherent fighting machine.

The Allied command on the Western Front was not in favour of a single American soldier being sent to Italy – even for what were primarily propaganda purposes, because, they argued, the War would in the last resort to be won or lost on the Western Front. What effect, then, was Allied propaganda having on the morale of the German soldier there?

The answer is, little or none until 1918 – or, to be more exact until the summer of 1918. Until then there was no significant deterioration in discipline, no noticeable crack in morale, no massive desertions. Allied propaganda over the previous years may have stimulated homesickness, disappointment at hopes deferred, disillusionment with the General Staff, concern about the fate of families at home, but there was nothing which could even by the wildest stretch of the imagination be described as a collapse in morale, or refusal to fight on.

In 1918 the situation altered drastically. Ludendorff had brought more than a million men across from the Eastern Front. He had launched his spring and summer offensives, taking advantage of the numerical superiority these million men gave him. He had made great advances – at great cost – but without achieving his strategic objectives. His offensives had been fought to a standstill, and now, in the summer of 1918, the Allies – reinforced by the Americans – were counter-attacking, and there was no telling where their advance would stop.

For the first time since August 1914 the German soldier was receptive to the message: Your cause is hopeless; your blood is being spilt in vain; you owe it to yourself and to your family to survive.

To stress the last point in this message, the British, the French and the Americans scattered leaflets over the German lines emphasizing how well prisoners-of-war were treated by them. This had been done by all the belligerent powers throughout the War to counteract propaganda that the enemy treated his prisoners abominably. It had had little effect – especially on the German soldiers – up to 1918 because up till then they had felt confident, or at least vaguely hopeful of victory. Its effect in 1918 was significantly greater. Although there were even in the summer and autumn of 1918 relatively few instances of desertions, German soldiers appeared readier than previously to allow their positions to be overrun and let themselves be taken prisoner. As a British military intelligence officer reported at the time, 'when I saw soldiers of a crack German regiment joking in one of our prisoners-of-war cages just behind the lines, and telling one another that they were glad to be out of it in one piece, I knew the end could not be far off'.

The Americans were undoubtedly the favourite enemy to be captured by. During Ludendorff's spring and summer offensives, the Germans had overrun a number of American positions. Their first surprise was to find American troops in the front-line at all. Their generals had told them that the Americans could not be in Europe in strength until 1919 – by when Germany would have won the War. Their second and more important surprise was their discovery of American army stores: white bread, kilos of butter and jam, meat, sacks of potatoes, endless packets of cigarettes and masses of matches.

More than one American army store was captured by the Germans in the course of Ludendorff's offensives, and the effect on the Germans was more telling than if their capture had been carefully planned by

George Creel in Washington, or Lord Northcliffe in Crewe House in London. George Creel's men in France learnt, through German prisoners of war, of the image of relative affluence and material superiority which the capture of these stores caused, and rushed to exploit this impression. They had printed a card which was an exact reproduction of the official German field postal card. On one side it said:

> 'Write the address of your family upon this card, and if you are captured by the Americans, give it to the first officer who questions you. He will make it his business to forward it in order that your family may be reassured concerning your situation.'

On the reverse side the postal card read:

> 'Do not worry about me. The war is over for me. I have good food. The American Army gives its prisoners the same food as its own soldiers: Beef, white bread, potatoes, beans, prunes, coffee, butter, tobacco, etc.'

American propaganda also stressed the hopelessness of Germany's continuing to fight. Creel's men had hundreds of thousands of leaflets dropped over the German lines and over Germany which took the simple form of just asking a series of questions:

> Will you ever again be as strong as you were in July 1918?
> Will your opponents grow daily stronger or weaker?
> Have your grievous losses suffered in 1918 brought you the victorious peace which your leaders promised you?
> Have you still a final hope of victory?
> Do you want to give up your life in a hopeless cause?

And to back up these questions, the Americans scattered masses of leaflets showing a row of soldiers whose size varied with the monthly increase in the number of American soldiers in France. 1 million 900,000 Americans are now in France, one such leaflet stated, and more than ten times as many stand ready in America.

The British, under Lord Northcliffe's direction at Crewe House, contrasted the hopes held out by Germany's leaders with reality:

The War will be over by Christmas 1914	:	Instead, stalemate
Unrestricted U-boat warfare will starve England into submission	:	Instead, English rations are higher than Germany's

Victory in the East will give Germany access to the grain of the Ukraine	:	Instead, it keeps enormous German armies in the East, and not one train load of grain has arrived
Britain and France will be knocked out of the war before America can intervene	:	Instead, 2 million American soldiers are already fighting next to their British and French comrades. By the end of the year there will be 5 million
One last offensive by Ludendorff will win final victory in France	:	Instead, Germany's armies are in retreat everywhere in the West

False hopes will not win the War

When the British, together with their Allies, went over to the offensive, they distributed maps over the German lines and Germany itself, upon which their gains were plainly marked. They also had distributed a leaflet which reprinted a statement alleged to have been taken from a German newspaper. It read:

'A few weeks ago it appeared as if our armies were very near their goal, the defeat of the enemy force, and peace. But what a change!'

Another pamphlet, alleged to emanate from a German source, carried the simple legend:

'Today we are in retreat. Next year we shall be destroyed.'

More disturbing still was a leaflet British aeroplanes dropped over German Naval ports. It gave details of the loss of 150 German U-boats and their commanders. Subsequent investigation showed that these leaflets caused immense depression among the naval families living in these ports. Their country's cause indeed seemed hopeless.

Still, all Britain's efforts at undermining German morale would have been of little use in 1918, the moment when they had the opportunity of achieving maximum impact, unless her engineers had not succeeded in devising a method of delivering the thousands of leaflets that were being printed. In the earlier years of the War Britain and her Allies had used aeroplanes, and aeroplanes were still being used especially to drop leaflets on Germany itself, but the number of leaflets an aeroplane could carry in those years was plainly limited. Manned balloons had been used, but again the space available in them was limited, and the crew had duties to perform other than dropping leaflets. Besides, manned balloons had to stay fairly close to their base behind their own lines and were therefore easy targets for enemy fighters. Various projectiles had been used on the Western Front, including trench mortars, but trench mortars were quickly abandoned. After all, it seemed inappropriate to use the same instrument for launching 'flying pigs' to maim and destroy as for disseminating propaganda advocating the justness of one's own cause and the iniquity of the enemy's. Fortunately by 1918 British engineers had designed an unmanned balloon which, according to the manner in which its special release mechanism was set, could scatter leaflets over distances ranging from ten miles to 150 miles. Equally fortunately the prevailing wind in the summer and autumn of 1918 was in the direction of the Allied advance. As a result, the British alone were able to drop more than 100,000 leaflets a day in August 1918. For the month of October the figure had risen to 5½ million.

The leaflets were concerned not only with pointing out the hopelessness of Germany's position. Their other main theme was that the German people should rid themselves of the Kaiser and the militarists who had led them into their present predicament, and that if they did, Germany too would eventually be able to look forward to a happy future as a member of the comity of nations. This second theme reflected the thinking of H. G. Wells, the head of the German department at Crewe House from February to July 1918. It was a theme that echoed President Wilson's attitude and formed a main part of George Creel's propaganda. The War, so the argument ran, was being fought not against the German people but against her rulers. Of course Germany would, to begin with, have to make some sacrifices, territorial and financial, but in the long run the future of her people would be brighter.

The German command was appalled by the avalanche of leaflets that descended on their troops and their people at home. They deplored particularly the call to their soldiers and their soldiers' families to get rid of the Kaiser and his Government and put in somebody else instead. This was to them akin to treason.

Their attitude is unmistakably reflected in the autobiography of their Commander-in-Chief, Field Marshal Paul von Hindenburg: 'Ill-humour and disappointment that the War seemed to have no end, in spite of all our victories, had ruined the character of many of our brave men. Dangers and hardship in the field, battle and turmoil, on top of which came the complaints from home about many real and some imaginary privations! All this gradually had a demoralizing effect, especially as no end seemed to be in sight. In the shower of pamphlets which was scattered by enemy airmen our adversaries said and wrote that they did not think so badly of us; that we must only be reasonable and perhaps here and there renounce something we had conquered. Then everything would soon be right again and we could live together in peace, in perpetual international peace. As regards peace within our own borders, new men and new Governments would see to that. What a blessing peace would be after all the fighting! There was, therefore, no point in continuing the struggle. Such was the purport of what our men read and said. The soldier thought it could not be all enemy lies, allowed it to poison his mind, and proceeded to poison the minds of others.'

The German command offered bonuses for the delivery of enemy leaflets, pictures and books. This rate was as follows: 3 marks for a leaflet or picture of which no copy had up to the time of delivery come into the possession of the authorities; 30 pfennigs for every copy of a leaflet or picture of which the authorities already possessed a copy; 5 marks for a book.

But it was too late to stop the Allied propaganda onslaught with a fistful of marks and a handful of pfennigs – or, for that matter, to halt the German army's retreat. On 9 November 1918 it was announced that the Kaiser had abdicated and on 11 November Germany signed the armistice terms at Compiegne.

CHAPTER 6

Between the two
World Wars:
Retrospect and
Prospect

Less than 21 years separated the signing of the armistice of the First
World War from the outbreak of the Second World War. Inevitably,
the propaganda conducted by the belligerent powers in the First World
War had its effects on policy and its consequences were felt in the
Second World War and in the period between the two Wars.

One consequence of the First World War was that many Germans
tried to convince themselves that British propaganda had been the cause
of their defeat. This started even before the First World War ended.
General von Hutier, Commander of the Sixth German Army, sent a
message to his troops, which read:

'The enemy begins to realize that we cannot be crushed by blockade,
superiority of numbers, or force of arms. He is, therefore, trying a
last resource. While engaging to the utmost of his military force, he
is racking his imagination for ruses, trickery, and other underhand
methods of which he is a past master, to induce in the minds of the
German people a doubt of their invincibility.'

This message was captured by the Allies when they occupied
General Hutier's headquarters in the course of their counter-offensive
in 1918.

Many German newspapers printed leaders and articles on similar
lines. F. Stossinger of the *Frankfurter Zeitung* described British propaganda as 'the most complicated and dangerous of all'. The *Rheinische-
Westfälische Zeitung* complimented the British Propaganda Department on its untiring efforts and added: 'Had we shown the same
activity in our Propaganda, perhaps many a thing would have been
different now. But in this, we regret to say, we were absolutely
unprepared but we hope that by now we have learnt differently.'

The *Deutsche Tageszeitung* tried to strike a prouder patriotic note:
'We Germans have a right to be proud of our General Staff. We have a
feeling that our enemies' General Staff cannot hold a candle to it, but

we also have a feeling that our enemies have a brilliant Propaganda General Staff, whereas we have none.'

These were comments written before the armistice, but editorials of a similar nature continued to be printed in German papers and magazines after the armistice. In the *Tägliche Rundschau* Arnold Rechberg wrote in July 1919: 'It cannot be doubted that Lord Northcliffe very substantially contributed to England's victory in the World War. His conduct of English propaganda during the War will some day find its place in history as a performance hardly to be surpassed. The Northcliffe propaganda during the War correctly estimated ... the character and intellectual peculiarities of the Germans.'

Throughout the 1920's German writers and political analysts – and to a lesser degree Austrians – flooded the market with books on propaganda, and especially British propaganda. The over-all impression left by these voluminous out-pourings was that Britain had discovered a new weapon of war against which the best trained troops, the most efficient guns and aeroplanes were useless.

No one subscribed to this thesis more fervently than General Ludendorff when he wrote in his *War Memories*:

'Lloyd George knew what he was doing when, after the close of the War, he gave Lord Northcliffe the thanks of England for the propaganda he had carried out – Lord Northcliffe was a master of mass-suggestion. The enemy's propaganda attacked us by transmitting reports and print from the neutral states on our frontier, especially Holland and Switzerland. It assailed us in the same way from Austria, and finally in our own country by using the air. It did this with such method and on such a scale that many people were no longer able to distinguish their own impressions from what the enemy propaganda told them. This propaganda was all the more effective in our case as we had to rely, not on the numbers, but on the quality of our battalions in prosecuting the War. The importance of numbers in war is incontestable. Without soldiers there can be no war. But numbers count only according to the spirit which animates them. As it is in the life of peoples, so it is also on the battlefield. We had fought against the world, and could continue to do so with good conscience so long as we were spiritually ready to endure the burden of war. So long as we were this, we had hope of Victory and refused to bow to the enemy's determination to annihilate us.'

Ludendorff's implied tribute may have given Lord Northcliffe and some of his colleagues at Crewe House considerable satisfaction, but could too much significance be attached to what Ludendorff wrote?

Generals throughout history have always shown themselves reluctant to accept personal responsibility for their defeats on the field of battle. If unsuccessful, their memoirs have been filled with excuses and explanations which are meant to leave one wondering whether the writer ever made any mistake in devising and executing his strategy on the field of battle.

Ludendorff, it appeared, was no exception – nor, incidentally, were some of the Italian generals after Caporetto. The collapse of the War in the East in 1917 provided the opportunity for German numerical superiority in the West for the first time since 1914. It was because of Ludendorff's insistence on harsh peace terms at Brest-Litovsk, as we have seen, that only 1 million German soldiers could be brought West instead of nearly 2 million. So Ludendorff gained numerical superiority but not the overwhelming numerical superiority which might have ensured success for his strategy. His spring and summer offensives of 1918 made substantial progress, but failed to attain their ultimate objectives. And then the Allies began their counter-offensives. Germany's military position had become hopeless. It was only then that German soldiers, sailors and civilians became receptive to Allied propaganda. The cause of Germany's military predicament was Ludendorff's strategy, and he used propaganda to cover up his professional mistakes as Chief of the General Staff. Allied propaganda merely exploited that predicament; it did not cause it. Germany's military strength was exhausted and her armies defeated. The sole question that remained to be settled, was how soon that fact would be acknowledged, and further useless bloodshed avoided. It is to the credit of the German generals at the end of the First World War, that they accepted the inevitable and insisted on an armistice, instead of allowing a retreat to be fought back across the whole of Germany right to the Wilhelmstrasse in Berlin, as their successors did in the Second World War.

Nevertheless, Allied – and particularly British – propaganda was a convenient way for many Germans besides Ludendorff to avoid facing up to the unpleasant truth that Germany had been militarily defeated in the First World War. It was certainly an argument that suited Hitler admirably. The 'stab in the back' myth was a convenient excuse for military incompetence. In *Mein Kampf* he was lavish in his praise of British propaganda, and he frequently claimed that he based his own methods on Britain's. If by that he meant the political propaganda methods he used – and with success – to win political power in Germany, his claim is certainly not borne out by the facts. Britain's war-time propaganda was entirely different in purpose and execution from the internal political propaganda carried out by Hitler and his followers inside Germany between the two World Wars.

Germany's political situation at that time was unique to that country. Here was a highly industrialized nation defeated in war. Tens of thousands of its former professional soldiers had to accept jobs which they considered below their dignity – if they were fortunate enough to find jobs at all, no matter how badly paid. Millions were unemployed. Inflation had ruined a large proportion of the middle class and blurred the dividing line between it and the working class. The more imaginative of the younger generation were, as in every age and every country, looking for some kind of idealism by which to lead their lives; instead of a helpful guiding hand they met only cynicism. Millions everywhere had their grievances and were looking for scapegoats.

It was an ideal situation for Hitler to exploit; and exploit it he did. The title he gave his movement – National Socialism – could provide a home for everyone in the whole social and political spectrum of Germany.

While many in Germany were over-lavish – for their own reasons – in praising the success of British propaganda during the First World War, the reaction to it in America was one of suspicion. This did not reveal itself fully until the outbreak of the Second World War, but then it came as somewhat of a shock in Britain. Many Americans felt – rightly or wrongly – that they had somehow been lured into the First World War, and certainly the many books and articles which those involved in Britain's efforts in courting the United States had written after the War, confirmed them in that impression. Perhaps it was true, after all, that the U.S. had entered the War because of America's immense economic stake in Britain's survival? Perhaps the whole business had been settled discreetly behind closed doors over cocktails and dinner between members of the political and social élites of the two countries?

There were enough Britons alive at the outbreak of the Second World War to remember how pleased they had been with the discreet way they had wooed America in contrast to Imperial Germany's clumsy methods, and they were unpleasantly surprised that their reputation for discretness should have engendered so much suspicion.

What they forgot was that shortly before the U.S. had entered the War in April 1917, even President Wilson, who was staunchly pro-Ally, had remarked to friends: 'Oh, if only there were another way out.' His mood was shared by many of his fellow-countrymen. It was a mood that had been obscured during the inter-war years, not least by documentaries made of news-reels which had concertinaed history by showing American troops marching down Broadway covered by ticker-tape and cheered on by enthusiastic crowds as they embarked for France. In fact, the mood of the country was one of fatalistic accept-ance. It lasted through May and June, and the nation was aroused only when President Wilson and George Creel set about telling the people the ideals America was fighting for.

This time there seemed to be no ideals. Moreover, had not British publicists like Sir Gilbert Parker stressed that wherever possible, they found it advisable to let Americans argue their case? Dorothy Thomp-son discovered just how much suspicion and distrust this revelation had stirred up when she addressed a meeting of the Veterans of Foreign Wars in Newark, New Jersey. At the end of her speech some members of the audience demanded that she be investigated on the grounds that she sounded like 'a British agent' and that if her words were to be believed 'we will be in the war'.

Yet the Americans were by no means pro-German or pro-Hitler. In fact, according to extensive public opinion polls the number of those who favoured Nazi-Germany grew smaller and smaller between 1939 and 1941. The organization called the Bund, led by Fritz Kuhn and financed by Goebbels, crudely repeated the 'blood calling to blood' call to German-Americans, as German propagandists had done in the First World War, and like the First World War propagandists, it failed lamentably. It not only antagonized the vast majority of German-Americans; it also offended the rest of America by its blatant support of Nazism. Moreover there were quite a number of home-grown Ameri-can organizations which in varying degrees supported Hitler's racialist theories: Father Coughlin's Christian Front, the Christian Mobilizers, the Paul Reveres, American Patriots, Inc., the Defenders, the Ku-Klux-Klan, the Silvershirts and the American Nationalist Party. Their extremism in the long term produced more enemies than friends.

Much more formidable opponents of the interventionists were genuinely isolationist organizations like the 'America First' movement, and no one challenged the patriotism and sincerity of dedicated isola-tionists like senators Nye, Borah or Robert La Follette. Equally, no one dreamt of accusing them of being pro-Nazi. Indeed, by the end of 1940 or the beginning of 1941, it had become a worse insult to call an American 'pro-Hitler' than it had been in 1916 to call a man 'pro-Kaiser'.

Indeed, according to a succession of public opinion polls, Americans were predominantly 'pro-Ally'. They were at the same time 'anti-war'. There were enough Americans to remember the rallying calls of American interventionists and of British propaganda. They seemed to be not dissimilar to those of 1939, 1940 and 1941: 'democracy', 'saving Western civilization', 'the independence of small nations'. But with memories of what happened in 1917 still fresh, the reaction now was: Do these calls mean that our boys will have to spill their blood in foreign lands?

The success of American interventionists and to some degree of British propaganda in the First World War proved its main handicap in the Second. There sprang up a propaganda movement against war propaganda. Eventually papers like the *New York Times* and the *New York Herald Tribune* and a host of private citizens, took up the fight against the anti-propagandists. 'It is insidious stuff', a writer said in the *New York Times,* 'which taken in too large doses, is likely to cause moral impotence and intellectual sterility.' Yet theirs was an uphill struggle. American public opinion was not to be shifted easily out of its predominantly 'pro-Ally, anti-war' mood.

It was against this background that President Franklin D. Roosevelt tried to give the Western Allies as much aid as quickly as he could. He exchanged 50 old American destroyers for the lease of British naval bases in the Caribbean. He gradually dismantled the Neutrality Acts piece by piece, always just winning sufficient Congressional support. When it came to Lend-Lease, he even secured substantial majorities in both the Senate and the House of Representatives. In July 1941, he had taken over from Britain the occupation of Iceland (then still formally linked to Denmark which was occupied by the Germans) and ordered the U.S. Navy to escort merchant ships destined for Britain as far as Iceland 'whenever such action seemed appropriate'; and in July 1941 he met Churchill in Argentia Harbour, Newfoundland, to draw up the Atlantic Charter setting out the Free World's plans for the future.

The occupation of Iceland and the order to American warships to escort merchant ships as far as Iceland was virtually a direct challenge to Hitler, and Admiral Raeder, the Commander-in-Chief of the German Navy, asked Hitler to allow him to instruct his U-boats to attack American warships since the United States was now virtually in the War. But Hitler, preoccupied with the Russian Campaign, refused to allow himself to be provoked by Roosevelt. He gave strict orders that U-boat commanders were not to attack American warships.

'Incidents', however, were bound to happen. On 30 October, 1941, the American destroyer *Reuben James* was torpedoed with the loss of 115 of her crew. The 'incident' would have sufficed in 1916 or 1917 – after the sinking of the *Lusitania* and the execution of Edith Cavell – to help propel the United States into a declaration of war. In 1941 the loss of the *Reuben James* met with a kind of resigned fatalism. President Roosevelt, who had asked Congress to approve the arming of American merchant ships, had his request passed by the Senate and the House of Representatives approved by majorities of less than 20 in November 1941. The support Congress gave him was in fact considerably less than his majorities in the two Chambers of Congress on Lend-Lease. It made clear to the President that if such small majorities were all he could secure on the comparatively minor issue of arming merchant ships after the torpedoeing of the U.S.S. *Reuben James,* he could not depend on Congress to vote through a declaration of war. It seemed as though Congress and American public opinion could not be shaken out of their 'pro-Ally, anti-war' attitude.

And then, less than a month later, on 7 December 1941, Japan attacked Pearl Harbor. On 11 December 1941 Hitler declared war on the United States and President Roosevelt asked Congress to vote on a motion declaring that a state of war now existed with Germany and Italy – as well as Japan, a motion Congress had previously approved. There was not one dissenting voice. The Congressional vote in both Houses was unanimous.

While American interventionist and British propaganda in the First World War certainly helped to delay United States intervention in the Second, there was another field of propaganda – and particularly British propaganda – which had its repercussions before and during the Second World War. It is most easily summarized under the heading of 'atrocity propaganda'. It was a form of propaganda in which all the belligerent powers indulged. It was a form, moreover, in which none – whether they were German, British, French, American, Canadian, Austrian, Russian, Serbian or Italian – were too scrupulous. None of the belligerent powers paid too meticulous attention to the truth. They all acted on the principle that people were on the whole prepared to believe the worst of the enemy. The gorier, the more revolting the story, the greater the likelihood of its being believed and strengthening the will to fight on or converting neutrals. The calculation behind the dissemination of atrocity propaganda was the same in all the belligerent countries involved in the First World War. It could hardly be called a very moral calculation. The difference between the belligerent powers was not one of morality but of effectiveness. And Britain's atrocity propaganda happened to be the most effective – at least while the First World War was being fought.

The problem of effective propaganda is, however, that it is likely to generate reactions long after its immediate, short-term aims have been achieved. That is exactly what happened to Britain's anti-German atrocity propaganda. Lord Bryce and H. A. L. Fisher accepted much of the evidence on alleged German atrocities in Belgium with which the lawyers attached to their Committee of Enquiry presented them, in the emotionally charged atmosphere of the First World War, and they stood by the findings of their Report after the War. Others in Britain found it more difficult to accept their findings and what the press made of these findings during the War, in the calmer atmosphere of the post-war years. Was every German – simply because he was a German – really a barbarian given to bayonetting babies, raping women and cutting off the hands of teen-age boys at the drop of a spiked helmet? During the 1930's Sir Harold Nicolson, later to be the author of the distinguished book on *George V and His Times,* expressed his misgivings in the House of Commons at the unscrupulous methods official British propaganda agencies used in spreading anti-German atrocity propaganda. He was not alone in his condemnation. The pity of it was that Sir Harold's condemnation came at a time when Hitler had won power in Germany and was beginning to practice the atrocities of which Germany had been accused in the First World War. In other words, the righteous indignation of Sir Harold Nicolson and many others like him in Britain at Britain's exploitation of atrocity propaganda in Germany coincided with Hitler making a reality of many of those alleged atrocities. So many people in Britain who could by no stretch of the imagination be accused of being pro-Nazi, tended to be rather sceptical of stories of illegal imprisonment, torture and summary execution in Nazi concentration camps. They had been fed all this propaganda before. There now appeared to be little basis for it then. So why believe it now?

Hitler went out of his way to foster this mood. Front-line soldiers who had been in the trenches – no matter on which side – had always known that these atrocity tales were nothing but the feverish fabrications of civilians sitting in safe luxury hundreds of miles behind the lines. Let the front-line soldiers get together; they could understand one another and tell truth from falsehood. Throughout the 1930's Hitler made time to meet Old Comrades' Associations visiting Germany, especially if they came from Britain or America.

The somewhat hysterical anti-German atrocity propaganda of the First World War affected even those who had not been in the trenches of Flanders and Northern France. It affected the whole climate of public opinion throughout the Western world. After all, one nation could not be as wicked as the Germans had been made out to be in the First World War. Because British atrocity propaganda had been so effective in the First World War, many people in Britain tended to be a little sceptical even of first-hand accounts of Nazi behaviour inside

Germany. Even Britons who had in person watched some of the anti-Jewish pogroms in Germany discovered that their reports were listened to with polite disbelief. The very success of British anti-German propaganda in the First World War proved a serious handicap in convincing people of the true nature of the threat and poisonous challenge with which Hitler and his régime were presenting the world.

Certainly this was true of the United States. Allan Nevins, the historian, wrote in the magazine section of the *New York Times* of 29 October 1939 – almost two months after the outbreak of the Second World War in Europe: 'Another Bryce Report would be read with so liberal a sprinkling of salt that few would stomach it. Stories of murdered hostages would be examined not emotionally but with a stern criticism of authorities; tales of crucified prisoners and bayonetted babes would be likely to arouse a revulsion of feeling. . . .'

Despite extensive press coverage and eye-witness accounts of what went on in Hitler's concentration camps, many people were not really prepared to accept fully what they had been and were being told. First World War propaganda had made them sceptical. It is for that reason that when the Allied Armies in 1945 overran the concentration camps in Germany, General Eisenhower asked members of the American Congress and the British Parliament to be flown out to see for themselves. The exaggerated and often fabricated atrocities of the First World War should, in his view, not be allowed to obscure the real atrocities of the Second.

Undoubtedly, then, propaganda between 1914–1918 affected propaganda between 1939–1945. And nowhere more so than in the field of communications. In the First World War the wireless had still been a comparatively inefficient and unreliable instrument. By 1939 the radio set had virtually become a standard piece of household equipment, and in the Second World War the radio became the main means of communications in the propaganda battle.

PART TWO Second World War
1939—45

CHAPTER 7 German Propaganda attacks with the Wehrmacht

In the autumn of 1939 the Germans showered the Maginot Line with leaflets in the shape of fallen leaves. They bore the inscription:

Automne
Les feuilles tombent
Nous tomberons comme elles
Les feuilles meurent parceque Dieu le veut
Mais nous, nous tombons parceque les
Anglais le veulent
Au printemps prochain personne ne se souviendra
des feuilles mortes ni des poilus tués.

'Autumn, the leaves are falling. We will fall like them. The leaves die by the will of God, but we fall by the will of the English. Next spring nobody will remember either the dead leaves or the dead soldiers.'

It was hardly a message likely to raise the morale of soldiers quartered in the underground system of fortifications known as the Maginot Line, waiting for and wondering what would happen next. They had been called to the colours shortly before the outbreak of the War. It was the second time they had been mobilized within 12 months. The first was in the autumn of 1938 during the Sudeten crisis. When Hitler, Mussolini, Chamberlain and Daladier had signed the Munich agreement, they had been stood down. Now they were back in uniform again, and this time war had actually come. They had not responded to the call to arms with any enthusiasm. In this they were not alone among the belligerent powers. Neither Paris nor London had in 1939 witnessed the wild, ecstatic popular demonstrations which had greeted the outbreak of hostilities in 1914. Nor for that matter had Berlin. Even there the predominant mood was one of resignation.

Yet there it mattered less than in Paris or in London. There even those who had doubts and misgivings knew that it was wiser to keep one's counsel. Moreover, the short, sharp Polish campaign was a convincing example of how the theory of the 'Blitzkrieg' should be executed in practice. Perhaps the Wehrmacht was an invincible instrument of war after all? Perhaps the propagandists' claims were correct?

Needless to say, the Nazi propagandists exploited this attitude. It was the way they had learnt to exploit propaganda in the 1920's and 1930's. Propaganda and various degrees of coercion, such as the use of storm troopers in street battles against demonstrators belonging to other political factions or against individuals, had been employed to win political control inside Germany. Both propaganda and coercion were then brought in to maintain control. They went hand in hand. Persuasion or propaganda and coercion had been used in harness as an instrument of social control inside Germany. They were now to be employed as an instrument to win control over Nazi Germany's enemies.

Consequently the German propaganda machine at the outbreak of war became to all intents and purposes a branch of Germany's armed forces. It was made to serve military objectives and was integrated into military operations just as were the Luftwaffe, the U-boats, the infantry, the Panzers or the paratroops.

As the Wehrmacht launched its campaigns in 1940, the dominant theme became the uselessness of resistance. Just four days before the attack on Norway, Oslo, for instance, was treated to a first showing of the film 'Baptism of Fire', an account of the previous September's campaign in Poland which showed how the Polish Army had been battered out of existence in under three weeks.

If at the outset of every campaign the accent was on terror, on paralyzing the opponent's will to resist through fear or helplessness, the tone in which the victims were addressed became somewhat more solicitous once the German High Command was able to claim its first decisive successes. Why go on, it asked, why should so many beautiful towns which were still unscarred by the horrors of war have to suffer the fate of Warsaw? Why should so many young men have to lose their lives in hopeless battles?

These broad themes of German propaganda occurred in every campaign in 1940, and they were applied in an infinite variety of ways, depending on local conditions and on the military situation prevailing at the time. They were thrown at the enemy by means of leaflets, films, placards, planted stories, megaphones and loud-hailers and by a new medium, the radio, which was used more extensively and intensively than any of the other means of communication in the Second World War because it enabled friend, foe and neutral to talk to and at one another across battlefields, oceans and frontiers.

Whatever the means of communication, German propaganda was not dedicated to disseminating the truth if it was thought an untruth would serve better. While Norway was being overrun in 1940, Goebbels called on Hitler one day to find him delighted at the news that a British cruiser had been sunk outside Trondheim. Ribbentrop, who happened to be present, remarked with a touch of pride in his voice that it had been he who had been the first to break the news of this major British naval setback to the Führer. Goebbels had some trouble in convincing Hitler that the news of the sinking was a fake planted by his propaganda people. The idea behind the plant, he explained, was either to lure the British into a denial that a British cruiser had been sunk off Trondheim – which might provide useful information for German naval intelligence that British cruisers were operating in Norwegian waters – or to induce the British to attack the truthfulness of news from German sources by saying that there were no British cruisers in Norwegian water at all – in which case, he, Goebbels, would be able to tell the Norwegians who were still resisting, that they really had a splendid ally in Britain which would not risk the sinking even of a cruiser to come to their assistance. Either way the planted fake news story could not but help the Wehrmacht in Norway. Hitler, though disappointed, saw the point. Ribbentrop's reaction is not on record.

A study of German broadcasts to Holland, Belgium and France in 1940 shows how Nazi propaganda operated as a branch of military operations. On 10 May 1940, the day the Germans invaded Holland, Belgium and Luxembourg, the German Radio informed its French listeners: 'The Luftwaffe now virtually dominates the North Sea. The successes up to now are only beginning. There will be very unpleasant surprises for Britain and France.'

On 11 May, the second day of the invasion, the Dutch were told: 'At daybreak this morning, German parachute troops were landed near Rheims. The object is to prepare the way for reinforcements, and by occupying key points to prevent the enemy from carrying through their movements. The French forces and the Maginot Line will thus be attacked in the rear.'

Having thus tried to convince the Dutch people of the hopelessness of their military situation, the Germans then appealed directly to the Dutch army: 'Soldiers, for whom are you fighting? For whom are you allowing yourselves to be butchered? For the capitalists in Holland, France and England?'

Even before the formal surrender of the Dutch Army, the Germans had already switched their main propaganda offensive to the next victim on their list, to Belgium. Broadcasting in Flemish, they directed their appeal to one half of the Belgian nation: 'Flemings. Soldiers! In the Belgian State you have always been citizens of inferior status. When you come over to us, we shall treat you as the sons of a kindred German nation.'

And broadcasting in French, the Germans addressed themselves to the other half of the Belgian nation; the Walloons: 'If you help the Germans now, they will also help you later to develop your country. Belgium doesn't care a damn for you. She only expects Walloon workers to die for her.'

And all the time propaganda, persuasion, worked hand in hand with coercion, with the alleged invincibility of the Wehrmacht. Of course, the sudden shock of massed tank and dive-bomber attacks must have had the effect of being overwhelmed, swept away and aside by a vastly superior military tidal wave. Of course the equipment of the Dutch, Belgian and French forces must have seemed greatly inferior to that of the Germans. Yet even when the Germans could not deploy their armour and superior numbers, their mere appearance coupled with the myth of their invincibility – a myth that seemed to have been accepted as reality – caused troops in strongly entrenched and well-fortified positions to abandon the safety of their defences and surrender. It happened on the Juliana Canal, the Meuse River, the Meuse-Scheldt Canal, the Albert Canal and the Brussels Canal. German advance parties consisting of no more than one or two dozen men found battalions surrender to them as soon as they were spotted, and in many cases the Allied troops had failed to remove the markings from mine fields and booby traps which would have made it impossible to take their defensive positions except at the cost of heavy casualties. The Germans had expected heavy losses in the Dutch and Belgian river and canal system; they had also expected to have several of their divisions tied down there for a week if not longer. They pushed right through it in a matter of days, and their casualties were negligible. The ordinary German soldier was beginning to believe in his own invincibility.

On 28 May, King Leopold of the Belgians ordered the Belgian Army to lay down its arms, and the Germans immediately turned the Belgian surrender into a weapon for use in the propaganda drive against the French: 'Now you have lost your Dutch and Belgian allies. You have lost a fifth of your effective troops and still more of your modern war material. If you want to continue the fight, it will mean horrible slaughter ending in the destruction of France and of the French people.'

The Germans were by then rapidly advancing on Paris. The French

Government had left the capital. Was Paris to be defended or declared an open city and surrendered? The Luftwaffe made one raid on the capital. The rest was left to the German propaganda apparatus: 'Your Government is one which no longer deserves any respect from you. In a cowardly manner, it has left Paris, although it proclaimed that it would defend the city stone by stone. Why expose the city of Paris, with all its memorials of a glorious past, to insensate destruction? If the criminal desire of Reynaud is carried out, Paris will be nothing but a burning ruin.'

On 13 June, the French surrendered Paris to the German High Command by radio, and what remained of French determination to carry on the fight crumbled. As Somerset Maugham put it in *Strictly Personal*: 'Their cause was lost when rather than see Paris destroyed as the Poles had the courage to see Warsaw destroyed, they abandoned it without a blow.'

But France had yet to sue formally for a cease-fire, and to speed that moment as the Panzer columns plunged deeper into France, German propaganda to the French adopted a new, more strident and menacing one: 'You need sleep and rest. Fresh German troops are continually subjecting you to exhausting battles. There is no respite. From now on every day that passes when you do not ask for peace is a sin against France. The burden of sin is getting very heavy. . . . Take care!'

On 17 June, Marshal Pétain asked for an armistice. The Battle of France was over. Hitler was in control of continental Western Europe. Arms, supported by propaganda, had won the day.

CHAPTER 8

German 'Black'
Propaganda:
The Traitor of Stuttgart

The period between the end of the Polish campaign in 1939 and the beginning of the German offensive in the spring of 1940 is frequently called the 'phoney' war. It was an expression that gained wide currency in France and Britain and no doubt reflected the sense of a lack of purpose prevalent at the time. It was not the way in which Germany's leaders regarded those months of apparent inactivity. To them it was a period of military preparation for the ambitious campaign they knew lay ahead and in the propaganda sphere for 'softening up' enemy morale.

Four of the countries which were to become victims of Nazi aggression in 1940 – Norway, Denmark, Holland and Belgium – were at that time of course still neutral, and the Germans were careful not to give them any hint of their plans. In their propaganda therefore they confined themselves to stressing the invincibility of the Wehrmacht and the hopelessness of Britain's and France's cause, adding, however, that despite its superiority Germany remained ready to talk and negotiate issues on which there were differences. The purpose of this approach was twofold: first to engage the sympathy of the many sincere men and women in the four then still neutral countries who were trying to use the prolonged lull in the fighting to mediate and find some kind of peace formula acceptable to both sides, and, secondly, to divert and still any suspicions some of the people in the four countries might be having about Germany's true intentions towards them, by a show of sweet reasonableness. For good measure they never failed, whenever opportunity offered, to refer in flattering terms, especially when addressing the Scandinavian countries, to their peoples as joint guardians and trustees of the great Aryan tradition. Probably only the home-grown Fascist movements treated these racialist references as a compliment.

Germany's principal target of its 'softening-up' propaganda during the 'phoney' war, however, was France. It was there that the ultimate success of the campaign Germany was planning for 1940 to win control of the Western European Continent, would be decided, and it was – first and foremost – France's will to resist which had to be undermined.

The most ingenious and probably the most effective venture in this sphere was a 'black' radio station which the Germans operated from the autumn of 1939 until the spring of 1940. 'Black' stations are stations which pretend to be what they are not. This one pretended to be a 'freedom' station operating from inside France. Its main speaker was a man called Ferdonnet who purported to be a patriotic Frenchman whose love of country had prompted him, at great risk to himself, to try to open the eyes of his fellow countrymen to the evils around them and so to save and re-build France before it was too late.

Of course the idea of 'freedom' stations was not new. Various opposition groups to Hitler had tried during the 1930's to operate some from inside the Reich or from countries bordering on the Reich. Their technical resources were limited, and they rarely reached anything approaching a substantial audience. Not only was their signal usually weak, but they had to keep constantly on the move so as to avoid detection by squads equipped with technical devices to track them down. Such squads operated not only in the Reich but also in the countries bordering on the Reich which had no wish to exacerbate relations with Hitler Germany. As a result of constantly having to be on the move, these 'freedom' stations could hardly ever stick to the announced times of their broadcasts – a circumstance which, combined with the weakness of their signal, militated against their building up not only a large but a regular audience. Needless to say, in the strict sense of the term, they were not 'black' stations because they said openly what they were and what they were after. They did not pretend to be what they were not.

Ferdonnet's station, by contrast, was 'black'. It merely exploited the idea of the genuine 'freedom' stations. It was a convenient disguise. It was a disguise his listeners saw through and no-one minded their seeing through it. Frenchmen listening to him knew that a clandestine radio station could not operate within France – and especially not in wartime – without being located by modern tracking devices in a very short time. So the all too transparent disguise wove a kind of conspiratorial bond between Ferdonnet and his listeners. Let us hear what Ferdonnet has to say, became the attitude among an ever-increasing number of soldiers in the Maginot Line as the long, cold, uncomfortable winter of 1939–40 wore on; he will be more entertaining, more provocative, more stimulating than the insipid stuff our own official radio is dishing out.

Though conceived in the tradition of the 'freedom' stations of the 1930's, Ferdonnet had a great advantage over them. What he said was backed by the formidable technical resources of Radio Stuttgart. His talks and commentaries reached out across the Rhine to the men in the Maginot Line and beyond deep into France loud and clear. What the 'traitor of Stuttgart' had to say could be picked up easily on any ordinary radio set.

But why did he operate under the guise of a 'black' station, a guise which fooled no-one? Why did he not give his commentaries within the framework of the 'White' Germany French-language broadcasts? Primarily because by using a 'black freedom' station as his means of communication, he appeared free from official constraints. As an 'official' broadcaster of the white German state radio, he would have labelled himself openly as a French renegade willing to do the Nazis' bidding. He would set up an immediate antipathy against his broadcasts. Except for Fascists among his audience he would have been a 'switch-off'. But the point about him was that he was not preaching Fascism. He was broadcasting under the colours of a French patriot. Though they were known to be false colours, his voice did not set off instant waves of revulsion. He may have been pro-German, – otherwise what was he doing in Stuttgart? – but at least what he said did not bear an official imprint; and during the winter of 1939–40 a great many Frenchmen developed a deep distaste of anything reeking of officialdom. Did he not even sometimes call the Germans 'Boches'? He was irreverent; his jargon was that of the ordinary French soldier when it suited him; he knew all the current jargon; above all, he knew about the grouses and grumbles of the *poilu,* the intrigues in Paris, the ways in which men were making money out of the war. He was the only one who seemed prepared to call a spade a spade. Neither the official 'white' French radio nor the official 'white' German radio appeared ready to do so. All they produced were official hand-outs.

Ferdonnet was able, intelligent and formidable in the way he presented his arguments. He seemed to possess a sixth sense for the mood of his audience, and the facts he produced tallied with his listeners' own experience. The winter of 1939–40 was exceptionally cold, and the combination of harsh frost on the surface, heating down below, insufficient ventilation and underground streams of which not enough account had been taken when the Maginot Line was constructed, produced an uncomfortably high degree of dampness in places. I wonder, Ferdonnet mused, if the men who constructed the Line belonged to the Navy – as the men in their fortifications watched the water trickle down the walls. Still, he added, someone somewhere made a lot of money out of it.

Claustrophobia, he said on another occasion, was a condition that afflicts even the healthiest of persons. That was one reason why submarines were usually given extra-long spells of leave after each tour. But was there a special leave allowance for the soldiers in the Maginot Line? Of course not. Your senior officers, he continued, may not get much more than you – on paper. But did it not seem a little odd how often they went off on staff conferences? He happened to know of one that had taken place in Arras recently. The morning, he told the *poilus* in the Maginot Line, had been spent at Headquarters. The conference then had adjourned to a restaurant, which he named and which was well-known throughout France for the excellence of its cuisine. He went in detail through a six-course lunch, not forgetting to mention the wines which accompanied each dish – and he made this broadcast at a time when he knew that many of the men listening to him were dunking their bread in their stews and washing their food down with their ration of rough red wine.

Ferdonnet poured forth a stream of talks and commentaries designed to play on the doubts and suspicions of his listeners, to set the French against the British, the French public against its government, workers against employers, poor against rich, the *poilu* in the Maginot Line against his General Staff.

Men working in French tax offices, he announced gravely on one occasion, were quietly being given exemption from military service unlike ordinary French citizens. The reason, he explained, was that they were needed to collect the money to sustain the war effort. Could ordinary soldiers not tell how necessary these tax collectors behind the front-line were when they collected their few miserable sous in pay. Not that that pay would get them very far, he added; even if it had been saved for weeks for their next leave. For if they went anywhere to Northern France on leave, they would find the cafés full of British soldiers with hundreds of francs to spend. They would not stand a chance against the much better paid British.

This broadcast was made when there was in fact considerable bitterness in some parts of northern France about the differences in pay between French and British soldiers.

Why, Ferdonnet asked, was France fighting Britain's war – the war of a crumbling, plutocratic, decadent empire? France was allowing herself to be shamelessly exploited, he claimed, and being treated little better than a British colony. France had mobilized millions of her young men, but British soldiers in France could be counted only in their thousands.

Ferdonnet repeated his taunt about the comparative smallness of the British contribution to the ground forces in France over and over again. It was during this period that the French service of the B.B.C. received a letter from a French farmer asking if a British soldier could not be sent out to take over the duties in the French army of his only son. He needed his son, he explained, because there would be so much work to do on his farm in the spring and summer.

Relations with Britain and the British were a theme to which Ferdonnet returned again and again. Were there, he enquired rhetorically, any British soldiers in the Maginot Line? Of course not – the British knew better than to entomb themselves. They lived in comfortable billets, their pay – as everyone knew – was better, and – sad to relate – there were some Frenchwomen who yielded to the persuasion of the Briton's bulging pocket book. It was altogether a grim situation for France. The only way out was a change in her leadership. There was a little time left yet, but not much.

Ferdonnet's real strength, the power of his attraction, lay in his incisive and repeated prodding of the sore points of France's body politic. He did not invent the doubts, the differences, the suspicions, the misgivings and prejudices to which he returned unceasingly in his broadcasts from Stuttgart. They were already there. He merely widened and deepened them. When he talked of 'the corrupt politicians of the Third Republic', or of 'the sinister machinations of the financiers', of 'the men who control the trusts', he was echoing what a great many loyal and patriotic Frenchmen were saying in their cafés – often in more pungent language than he himself employed.

Ferdonnet was a master of his craft. He had studied his audience. He had the technical means of getting across to it, and he enjoyed a high degree of credibility. This material, however, was not of the kind on which empires are built but by which they can, in suitable conditions, be sped on their way towards disruption and destruction.

CHAPTER 9

The Allied Response

Against the German propaganda onslaught at the beginning of the War the Allies had singularly little to offer. Bombers of the Royal Air Force took off for Germany, but their cargo was not bombs but leaflets. A story made the rounds in London during the first winter of the black-out of a naval officer warning an R.A.F. pilot friend of his not to drop the bundles of leaflets without first unwrapping them. 'After all, you don't want to hurt anyone do you?'

Another story told of an R.A.F. pilot who returned to his base 24 hours late. When asked to explain why it had taken him so long to fly to Germany, drop his leaflets and return, he raised his eyebrows in shocked embarrassment. 'Drop my leaflets, sir! Good heavens, I would never have dreamed of doing that. I landed in Germany and went from house to house, slipping one under every door.'

There is little evidence to suggest that leaflets had much effect – although they stung Hitler into saying in the course of a speech in the Market Square in Danzig: 'They must consider the German people as stupid as themselves. . . . If they want to know something about propaganda, let them learn from us.' And Goebbels described English propaganda as 'childish and laughable; it does not in the least disturb our sovereign feeling of security.' It appears that his 'sovereign feeling of security' and that of his fellow Nazi leaders was not enhanced by an R.A.F. leaflet giving details of the not insignificant amounts of money leading Nazis had – according to the American Press – salted away outside the Reich in foreign lands against the eventuality of a rainy day. According to that leaflet Goebbels put aside some 35,960,000 Reichmarks in places as far away as Buenos Aires and Osaka – almost 6 million Reichmarks more than Goering.

But these were pinpricks. On the whole, the Allied effort in the field of information and propaganda was unimpressive. Where German propaganda was sharp, purposeful and precise, Allied propaganda often seemed vague, uncertain and feeble. But then, was this the fault of those whose job it was to sustain the morale of the Allies and undermine that of the enemy?

Propagandists are not prime ministers and cannot usurp the functions of cabinets; they cannot lay down policies and issue directives to implement them. They can operate only within the framework of the aims and purposes laid down by their countries' leaders. And during the period of the 'phoney war' there was no such framework.

Admittedly, the governments in London and Paris knew what they were fighting against. They were against Hitler and the 'isms' he stood for, but they gave no hint of knowing what they would do once they had removed him, of the kind of Europe they wanted to see. Nor was there any indication of how he was to be removed. Were Britain and France going to fight an aggressive or a defensive war? Was Germany to be starved into submission by blockade while her armies hurled themselves vainly against the West's impregnable defences? Or were the Allied armies to crash through the Siegfried Line? All was vagueness and uncertainty.

In those circumstances the propagandists and those responsible for the dissemination of news and information had little chance to make a profound impact. In Britain the B.B.C. used the period to build up and develop its broadcasts to friend, foe and neutral, and in their native tongues. The main ingredient of these broadcasts was objective news and comment. It may not have seemed a very exciting venture at the time, but after the *débâcle* of 1940, when Britain found herself alone, it provided the basis for the attempt to restore the morale of Continental Europe.

In France those concerned with maintaining morale at home and presenting the French case abroad seemed to be completely hypnotized by what the Germans were doing. They were permanently on the defensive, always trying desperately to nail some German propaganda point and often merely succeeding in giving it wider publicity by broadcasting or publishing news items or commentaries beginning with the phrase '*Il n'est pas vrai que . . .*' or '*Il est inexact que . . .*'. And as an alternative weapon they employed the crudest and most obvious forms of censorship and suppression.

Throughout France the clumsy censorship generated so much distrust that many Frenchmen had become devoted listeners to the B.B.C.'s French Service well before the German offensive in 1940. Not that London entirely escaped the charge of over-optimism, of not in every instance adhering to its own high standards of truth and objectivity when caution would have been wiser, or more important, more in accordance with the true situation. During the fighting in Norway many Norwegians – rightly or wrongly – felt that the B.B.C. had led them to expect great things of the British Expeditionary Force, and when the British failed to turn back or even stem the German tide, they were bitterly disillusioned. The remark of the then Prime Minister, Mr. Neville Chamberlain, that 'Hitler had missed the bus' became a sour joke, but it was a remark based on ignorance of what was going to happen and not on a wish to conceal what had in fact happened. Few in Britain at that time had a clear idea of what the country was up against. By 17 June there were no doubts on that score. It was a lesson bitterly learnt, and it was not forgotten in the seemingly interminable succession of long months of uphill struggle that lay ahead.

Words could not unmake the military disasters of 1940, and London, to its credit, did not try. The message broadcast in dozens of languages was simply: 'Germany won't win.' It carried conviction in Britain. The problem was: could it be made to carry conviction on the Continent?

The majority of Germans, especially those under 30 who had no personal memory of the First World War, no direct experience of how quickly disaster can follow triumph in war, lived in a cocoon of euphoric triumph at the time. They were impervious to London's message and, if aware of it at all, regarded it either as a joke or as a piece of incredible impertinence. The first reaction of most of occupied Europe was complex and involved. The shock had been too great. The agony of defeat, the humiliation of occupation tasted bitter, and many people, exhausted, separated from families and friends and drained of all emotion, did not know where to turn or what to think. Could London's simple message help to restore their confidence?

As the months rolled by, three factors combined to work in London's favour: First and foremost, there was the urge throughout occupied Europe to be rid of the Germans as quickly as possible – an urge that had lain buried for a while under the shock of defeat, the longing for a quiet, peaceful life at almost any price, but which soon asserted itself. And when it did, people naturally turned to London as the only active centre of resistance to Hitler.

Second, London's voice gained in respect and authority as soon as the Luftwaffe began its bombing raids on the British capital. Advice from a safe sanctuary – especially by a former comrade-in-arms who had for the time being escaped the general *débâcle* – would have been resented, but those on the Continent who listened to the B.B.C. felt they were being addressed by men and women in the frontline.

And third, there was the failure of the German propaganda machine to win the hearts and minds of any substantial section of the population in the countries occupied by the Wehrmacht. The fault, of course, lay not in Germany's propaganda techniques but in National Socialism itself. Once victory had been won, it revealed itself as devoid of ideas and ideals. The 'New Order' was a fine label, but a label for what? There was nothing to propagate – no vision of Europe to fire men's imagination, no philosophy to bind together the Old World's many divergent races and cultures. National Socialism had created a highly efficient war machine which knew how to conquer, but did not know what to do with its conquests. The war machine was served by a propaganda apparatus which knew how to intimidate, to confuse and to disrupt – but once the battle was over and victory won it had nothing to say.

Throughout the winter of 1940–1 the B.B.C. worked at nursing occupied Europe out of its state of shock and at rekindling hope and the spirit of resistance. The exploits of the R.A.F. helped. So did America's decision to 'Lend-Lease' Britain essential war material, a decision that meant that the British war effort was going to be backed by America's vast industrial power.

The problem in that dark, grim winter was to devise a sign, a symbol which crystallized the hopes and aspirations of the people of occupied Europe, rallied those who felt bold enough to resist the German occupation at least passively, and at the same time to undermine German confidence by showing them the extent of the opposition to them. In the V sign the B.B.C. found such a symbol. The letter V is probably the easiest and quickest to chalk up on walls or pavements. The morse signal for it (. . . –) is distinctive and can be tapped out without difficulty. Moreover, it was the rhythmic theme of many pieces of music, above all, of Beethoven's Fifth Symphony, whose majestic opening chords hammer out the rhythm and whose opening theme Beethoven declared to be 'Fate knocking at the door'.

The V for Victory campaign was launched early in 1941. It began with a broadcast from Radio Belgique in London: 'You should use the letter V as a rallying sign', it told the Flemings and the Walloons, 'because V stands for 'Victoire' in French and 'Vryheid' in Flemish'. Broadcasts to other countries followed the example of Radio Belgique, and the V sign was made to fit even in languages where V was not the first letters of the words that stand for victory or freedom. Listeners in Czechoslovakia, for instance, were reminded that V recalled John Huss's famous words 'Pravda Vitezi' ('Truth shall prevail') which Masaryk, the founder of the Republic, had adopted as the Presidential motto.

The response to the campaign throughout occupied Europe exceeded all expectations. The walls and pavements of whole towns would come out overnight in a rash of Vs. 'In France', a Swiss newspaper reported, 'The Vs seem to have come out on top.' Throughout the spring and summer of 1941 the campaign gathered momentum. At first the Germans tried to ignore the campaign, but by July they could no longer afford to do. Their armies had by then invaded the Soviet Union, and in support of the military onslaught on Russia their propaganda machine was trying to present their invasion as a sacred European crusade against the evil Asiatic, Jewish-inspired forces of barbaric Bolshevism. This was hardly the time when the greater part of the European Continent, which was supposed to be rallying to the holy cause of the crusade, could be allowed to be seen to come out in a rash of V signs, directed against the crusading armies and their leaders. Unexpectedly the German propaganda machine found itself, for the first time in the War, on the defensive. Something had to be done – and quickly.

The idea the Germans came up with, was to take over the V for Victory campaign, to claim that it was actually conceived by them and not by London. Plainly there must have been a classical scholar among their propagandists, because on his instructions occupied Europe was told that V did not stand for 'victory', 'Victoire' or 'Vryheid' but for 'Viktoria'. There was no time to check whether the ancient Romans in fact spelled 'Viktoria' as claimed by the propagandists but Hans Fritsche, one of Nazi Germany's ablest and most adaptable political commentators, went full steam ahead immediately on that basis.

On 10 July 1941 he urged the people of Europe to unite even more closely than they had been doing, against England under the 'V for Viktoria' sign. On 12 July he promulgated a European 'Monroe Doctrine' as a moral justification for the crusade against the barbaric, Asiatic Soviet Union as well as against non-Continental England and its distant transatlantic supporter, plutocratic America. On 15 July he elaborated on both these themes by praising the volunteers who had already come forward from outside Germany to help in the crusade against Asiatic Russia and at the same time by calling for more such volunteers. On 17 July he told his listeners that 'V for Viktoria' provided an appropriate symbol for the common cultural ideals of the nations of Europe, and said how fortunate Europe was to be given this rousing rallying call. It had all been due, he conceded, to a telegraphist who, anxious to transmit a special announcement about a military victory by the forces of European enlightenment over barbaric Bolshevism, had in his excitement simply tapped out 'V' which everybody knew could mean only 'Viktoria'. All who were fighting for Europe as well as all European radio stations, Fritsche continued, had spontaneously adopted the idea of the excited, enthusiastic telegraphist. On 20 July he accused Churchill of stealing the idea and pretending that he had invented the whole thing. The nerve of the man, Fritsche complained bitterly. He knew that Europe was already united under the 'V for Viktoria' sign; Churchill's claim was really an insult to the people of Europe.

Despite all the energy and ingenuity devoted to this campaign, Fritsche's efforts made no impact whatever. They were based on a completely false analysis of the mood and temper of the target audience. The B.B.C.'s campaign, by contrast, was not.

The Germans knew their propaganda campaign had failed when German official notices in occupied Europe and even notice boards in factories inside Germany using forced foreign labour, were so smothered in V signs and . . . — morse signals as to become unreadable. Not even the most starry-eyed German propagandist could interpret this as support for Fritsche's crusade, and so, propaganda having failed, the Germans fell back on coercion. They warned the populations of the occupied territories against being taken in by 'saboteurs and agents provocateurs in the pay of Britain'. When that had no effect, they resorted to cruder methods. They told householders that they would be held responsible for any V signs on their property or near their property.

London, in launching the 'V for Victory' campaign, was very conscious of the dangers it might entail, of the sudden vicious retaliation it might provoke on the part of the occupying power, and so in its broadcasts throughout the spring and summer it counselled caution. Resist by all means, was its oft-repeated message to the 'V-Army'; walk out of cafés when Germans enter; hoard copper and nickel coins because copper and nickel are important to the German war effort; go slow at work – but do not do any of these things if you are likely to be caught and arrested; do not expose yourself to German vengeance; be patient; this is not the moment for unnecessary personal risks.

The reason for this caution was of course partly humanitarian. London knew that it was in no position at that stage to bring any help or comfort to those who got into trouble, and that it would therefore have been wrong to ask them to make such a sacrifice. There was also a severely practical consideration. Britain – in early 1941 – stood alone, and on even the most optimistic interpretation of the course of the War in the spring of that year it seemed likely that it would be some years before Britain and such other nations as might rally to her side, could land on the Continent and roll back the Germans. In these circumstances it would have been playing into the hands of the Germans to

... *Ich Dien,* they say in Britain

... the war with Germany is a

Crusade to maintain democratic

liberty. But God help any

Englishman who dares to criti-

cise Churchill or his Government.

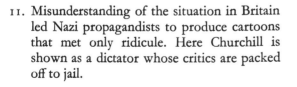

11. Misunderstanding of the situation in Britain led Nazi propagandists to produce cartoons that met only ridicule. Here Churchill is shown as a dictator whose critics are packed off to jail.

Les Otages

DECLARATION DU
Président Roosevelt
SUR LES EXECUTIONS
D'OTAGES EN FRANCE

Maison Blanche, Washington
25 octobre **1941**

" La pratique consistant à exécuter en masse d'innocents otages en représailles d'attaques isolées contre des Allemands dans les pays provisoirement placés sous la botte nazie révolte un monde pourtant déjà endurci aux souffrances et aux brutalités.

" Les peuples civilisés ont depuis longtemps adopté le principe qu'aucun homme ne doit être puni pour les actes d'un autre homme. Incapables d'appréhender les personnes ayant pris part à ces attaques, les nazis, selon leurs méthodes caractéristiques, égorgent cinquante ou cent personnes innocentes.

" Ceux qui voudraient " collaborer " avec Hitler, ou qui voudraient chercher à l'apaiser, ne peuvent point ignorer cet effroyable avertissement.

" Les nazis auraient pu apprendre de la dernière guerre l'impossibilité de briser le courage des hommes par la terreur. Au contraire, ils développent leur " lebensraum " et leur " ordre nouveau " en s'enfonçant plus bas qu'ils n'avaient eux-mêmes jamais été dans un abîme de cruauté.

" Ce sont là les actes d'hommes désespérés qui savent au fond de leur cœur qu'ils ne peuvent pas vaincre. Le terrorisme n'apportera jamais la paix en Europe. Il ne fait que semer les germes d'une haine qui, un jour, amènera un terrible châtiment."

Franklin D. Roosevelt

12. The USA was still neutral when Roosevelt spoke out against the execution of hostages in occupied France. Many Germans, even Nazis, were disturbed by his scathing condemnation.

allow the most active, resourceful and courageous men and women in occupied Europe, the spearhead, the potential leaders of the resistance movements which were then just beginning to take form, to expose themselves and so be caught by the Germans.

The success of the V campaign did not, however, alter the harsh reality of German military superiority on the Continent. With the exception of the Battle of Britain the German war machine had suffered no setback, no serious loss in men or material. Its morale was high, and in the spring of 1941 it again showed its formidable, ruthless punch and power. Within a matter of months, Yugoslavia, Greece, and then by aerial assault, Crete fell to it. British forces rushed to Greece's aid and fought hard to hold Crete. But as in Norway, they landed only to have to withdraw. Britain, standing alone, and with her stretched resources, was still no match for the Wehrmacht on the Continent.

It was a cruel demonstration of the military realities of the situation in 1941 and it caused deep distress in occupied Europe. Many of the letters and messages that reached the B.B.C. at that time were full of disappointment, bordering sometimes on despair. You in England, the letters said, begin things so well, but then you cannot follow through; you fail and everyone is left worse off than before – why do you attempt things when you cannot succeed?

Under such pressure the temptation in London to embroider on and exaggerate the importance of comparatively minor successes while playing down the significance of relatively major reserves, was great. It was a temptation that was resisted. Fortunately so, because to have yielded to it would have eroded the B.B.C.'s credibility sooner or later.

Instead, London through the B.B.C. stuck to the message: 'Hold on. We never promised victory this year, or next year, or the year after. But victory there will be eventually.'

The impact of London's message in 1940 and 1941 that 'Germany won't win' was considerably strengthened by what the Big Two Neutrals – America and Russia – were saying to occupied Europe. Without their orchestration of London's theme, often muted but seldom obscure, morale on the Continent might well have taken longer to recover.

Certainly many Frenchmen after their country's collapse needed more than Britain's assurance that everything was not lost. After all, Britain was the last remaining belligerent power facing Germany, and their own Government, while still a belligerent, had shown themselves notoriously adept in concealing grim reality behind a curtain of soothing words. Why should Britain behave differently? Only a neutral, a non-involved nation was likely to tell the truth, and they naturally turned to America, to the short-wave stations some of the big commercial networks in the United States had built up to reach across the Atlantic into Europe, in some cases with the help of money subscribed by first or second generation Americans from the countries now occupied by Hitler. These stations, with accounts by 'neutral' American correspondents direct from London – men like Edward Murrow and Quentin Reynolds – of Britain's mood and war effort often carried more weight than the B.B.C. Moreover, the cumulative effect of these broadcasts was to leave listeners in occupied Europe in no doubt where America's sympathies lay, and many concluded that Britain's chances could not be all that hopeless if she enjoyed the backing of America.

That impression was strengthened in the summer of 1941 when President Roosevelt, though the head of state of a country which was then still formally neutral, set up an Office of the Coordinator of Information. As Coordinator he appointed Colonel (later General) William J. Donovan, later to become head of the O.S.S. (the Office of Strategic Services) – one of the main branches of psychological and sabotage warfare once the U.S. had entered the War. An integral part of Col. Donovan's Office was an organization known as the Foreign Information Service (FIS). Its head became Robert Sherwood, the dramatist, and one of its main functions was to broadcast and to disseminate by other means news, the advantages of democracy and U.S. objectives throughout the world with the exception of Latin America, for which a separate agency was created. For people in occupied Europe, the work of Robert Sherwood's organization had only one meaning: the United States of America will be with us soon – and actively.

Equally important for the morale of many people in occupied Europe was the voice of neutral, non-belligerent Moscow. They remembered how the Communist parties in Western Europe had been encouraged to disrupt the Allied war effort in the winter of 1939–40; because was it not an imperialist war? They remembered, mostly with distaste, the Soviet Union's attack on Finland. And they remembered how Radio Moscow had dropped its violent anti-Nazi propaganda after Ribbentrop and Molotov had signed the Nazi-Soviet pact in August 1939. But was there not a subtle change in tone in Radio Moscow's broadcasts after the fall of France? Radio Moscow fully reported and repeated Churchill's declared intention to fight on. Some cynics on the Continent suggested that perhaps the men in the Kremlin may have been disturbed by the fact that Soviet losses in the limited winter campaign against tiny Finland were many times greater than Germany's in conquering the whole of Continental Western Europe, but whatever the reason, even the cynics conceded that it was remarkable that men as devoted to the ruthless application of Realpolitik should now go out of their way to make it clear that they did not accept the inevitability of a German *'Endsieg'*. The British, the Soviet Radio pointed out, could be tough and remarkably stubborn, their navy had ruled the seas for many generations and never been defeated, and – strangest of all in broadcasts coming from such a quarter – there were the vast resources of the British Empire to fall back on.

Not content with observations of a general nature, Radio Moscow broadcast reports by Tass correspondents on the bombing of London by the Luftwaffe. The bombing of Polish cities or of Rotterdam had been given no mention whatever. The London reports spoke of gigantic fires, of casualties, of evacuees, of there not being shelters. They also spoke of the cheerfulness of the people, of how they shared their food, their blankets, their cigarettes. They mentioned that many German bombs had been dropped on the poorer quarters, the docks, the East End. They also mentioned that Buckingham Palace had been bombed. And one Tass correspondent, after spending the night with an anti-aircraft battery in London, reported: 'The present system of anti-aircraft defences in England is much more impressive than anything the Luftwaffe has yet encountered.'

It was all put across in a detached, objective, matter-of-fact way, carefully calculated not to cause offence to the Germans. Its purpose of course was not to let hope die in occupied Europe, for this would plainly not have been in Russia's interests – and as it happened, it was also not in Britain's interests. And so non-belligerent Moscow helped to underscore embattled London's theme.

But the view of a non-belligerent Moscow, undermining German claims to invincibility, were one thing; the views of a Moscow at war with Germany quite another. The Nazi attack on Russia in June 1941 opened not only new opportunities but also cruel dilemmas in the fight for the support and allegiance of occupied Europe.

In many middle-aged and middle-class people Russia's entry into the War induced almost a split-mind condition. They had been brought up on and believed in the 'Communist bogey', the 'Red Menace', and while they recognized that they needed powerful outside help to get rid of the Germans, did that help really have to come from Russia? Surely there was nothing to choose between Nazi Germany and the Soviet Union?

Particularly cruel was the dilemma of the Poles and others in Central and Eastern Europe whose nations had for centuries lived uneasily between Russia and the German-speaking empires and had known occupation and repression by both sides. They felt caught between the Devil and the Deep Blue Sea.

London at that critical moment refused to vacillate. Its message was sharp and clear: 'Nazi Germany is the enemy – whoever fights Nazi Germany is our ally'. There was no equivocation, no hesitation, and the message struck home. Hitler failed to split those who were fighting him, by attempting to turn his invasion of the Soviet Union into a crusade.

CHAPTER 10

The Propaganda War
between Germany and
the Soviet Union

When Hitler invaded the Soviet Union in June 1941, he launched a war between two gigantic military machines which remained locked in combat for almost four years and which fought battles on a scale never before seen in history. Yet both sides entered this unique, colossal struggle virtually unprepared on the propaganda front, despite the fact that both were ruled by régimes which attached more importance to propaganda than any other kind of government elsewhere in the world and regarded propaganda as an essential ingredient in all forms of policy formulation and execution.

Hitler's mistake was worse than Stalin's and contributed to his ultimate defeat. It also helped Stalin to overcome his own initial mistake.

In the case of each of the countries Hitler had invaded and overrun in Western Europe, he had a detailed analysis made of the fissures, the grievances, the resentments and disputes which were weakening the structure of society in his intended victim's country, and his propaganda was tailored to extract the greatest possible advantage out of that analysis. In the case of Russia, no analysis of any sort was prepared. On the basis of his violent anti-Bolshevik rhetoric – which had stopped when early in 1939 the Soviet Union and Nazi Germany had agreed secretly not to attack one another publicly in their press and radio, an act of restraint confirmed by the conclusion of the Nazi-Soviet non-aggression pact of August 1939 – Hitler considered himself an expert on Communism and the Soviet Union and ignored the advice of Count von der Schulenburg, Germany's ambassador in Moscow until the outbreak of hostilities, and of Koestring, the German military attaché there. Both men pointed out that the Soviet Union was an ideal target for psychological warfare. The morale of the Red Army's officer corps was low after Stalin's purges in 1937 and 1938 in which Marshal Tukhachevsky and, so it was estimated, as many as 15,000 officers – most of them of senior rank – had been liquidated. Many others had been peremptorily retired, including promising officers such as Rokossovsky and Goronov – both of whom Stalin had to bring out of retirement during the War and both of whom rose to become Marshals of the Soviet Union. After the War, Rokossovsky, in fact, became the first Supreme Commander of the Warsaw Pact Powers' armed forces. Furthermore, the top Soviet leaders, Schulenberg and Koestring reported, were divided among themselves. None knew whom Stalin would purge next. In such conditions more than one of them felt that it was impossible to carry out a consistent policy on any issue vital to Russia and might therefore be prepared to make some accommodation in return for ending Stalin's reign of terror. The peasants were disaffected by Stalin's brutal collectivization of farms programme, and national minorities like the Ukrainians, the citizens of the former Baltic republics annexed by Stalin in 1940, the Azerbaidjanis, the Turkestanis, the Cossacks and the Armenians resented Greater Russian overlordship and oppression.

Hitler brushed all this information aside and described Schulenburg and Koestring as 'the two worst informed men on Russia'. To him all Russians and all the members of national minorities living within Russia's borders were murderous, barbarian, oriental. They were all *Untermenschen,* sub-human. It was a waste of time and trouble for any cultured, civilized Aryan to have any dealings with them.

He did not sense that he might be mistaken and that Schulenburg and Koestring might be right, even when in the summer and autumn of 1941 whole battalions and regiments and later complete divisions surrendered to the Germans. Russian officers, when captured, made no secret of their contempt for Stalin. He had purged the Red Army, they told their German interrogators, of their best officers at a time when war was imminent. Who was left to lead them now? Old Revolutionary War heroes like Marshal Voroshilov and Marshal Budjenny, both former sergeants in the Czarist cavalry who had never commanded anything larger than a platoon and who had learnt only one thing from their Czarist officers – incompetence. And now these two were put in charge of vast army groups. Soviet generals, colonels, majors and captains insisted to their German interrogators that they were soldiers who would not allow their men to be senselessly slaughtered as Czarist officers had done in the First World War. They also insisted that they were Russian patriots, and as Russian patriots many of them offered to take up arms against the Stalinist regime at the side of the Germans. Their offers of help were, on Hitler's instructions, firmly refused.

Naturally many German officers were surprised by the attitude of Russian soldiers and officers. They were equally surprised by the attitude of the civilian population. In Latvia, Lithuania and Estonia they were greeted as liberators, and the men begged to be allowed to join up to fight the Soviet oppressor. The Nazi answer was a cold refusal. The Ukrainians, the Cossacks and even many Greater Russian peasant families – Great Russians constituted the largest ethnic group in Russia – welcomed the German invaders with similar offers, only to be refused as curtly and rudely. Indeed, while the women were still handing bunches of flowers to embarrassed German soldiers whom they regarded as 'liberators', special S.S. task forces were busy rounding up their menfolk to transport them back to Germany to work on farms and in factories.

The Nazis proved completely insensitive to the opportunities with which they were presented so unexpectedly – opportunities which were far greater than any that ever existed in Western Europe. The Russians to them were not people with whom one could deal. The Nazis were prisoners of their arrogant conviction that the Russians were sub-human.

It was this conviction which the Nazis felt justified the plans they had for the future of Russia. These plans included the outright annexation of the Ukraine, White Russia, the Baltic States, the Crimea and the Caucasus. Their Slav populations were, to begin with, to serve as a cheap labour pool, the educated classes were to be liquidated, and eventually most of the people who had settled there were to be left to die through lack of medical attention and sanitation, and the vacuum thus created was to be filled by German colonists – 10 million in the first 10 years, and 100 million eventually.

Needless to say, such a policy made it impossible to reach any kind of accommodation with Russia's disgruntled soldiers, peasants or national minorities. That did not stop the Germans dropping leaflets on and broadcasting messages to the Soviet lines, announcing that they would liberate the Russians from their oppressive régime, that prisoners and defectors would be treated fairly, that the collectivization of farms would be stopped, and private property and trading restored. But as the fighting in Russia dragged on into its first winter and then into spring and its second summer, the gap between what the Germans were saying and what they were doing was becoming too big.

By the end of 1941 the Germans held 5 million Russian prisoners-of-war, two-thirds of whom, it has been estimated, were prepared to take up arms against Stalin if accorded the right conditions. Their offer was turned down. Instead, they were systematically starved – a first step in the programme to provide *Lebensraum* for future German settlers in Russia. A German army order read that 'the Russian soldier loses all claim to treatment as an honourable soldier according to the Geneva Convention'. Hundreds of thousands of civilians were shipped to Germany as slave labour, and those who were unsuitable, were liquidated by special S.S. units – a second step in the *Lebensraum* programme.

The gap between propaganda promises and reality became a chasm. Hitler, blinded by his prejudice against sub-human Slavs, may have felt that they lacked the perception, the sensitivity to notice. He was wrong. They did. His propaganda lost credibility and then turned into hatred against those responsible for it.

Where Hitler was obtuse, Stalin showed himself astute. His purges, his harsh policies had antagonized many and done great damage to Russian morale. Moreover, the Russian people were quite unprepared for the German onslaught. On 14 June – a week before the German attack when Soviet troops in forward positions could clearly hear the rumble of masses of German tanks taking up battle stations – he had authorized Tass to issue a communiqué specifically denying rumours of an early war between Germany and the Soviet Union. Tass said:

> 'All this is nothing but clumsy propaganda by forces hostile to the U.S.S.R. and Germany and interested in an extension of the war.
>
> Tass is authorized to state: (1) Germany has not made any claims on the U.S.S.R. and is not offering any new and closer understanding; there have been no such talks. (2) According to Soviet information, Germany is also unswervingly observing the conditions of the Soviet-German Non-Aggression Pact, just as the U.S.S.R. is doing. Therefore, in the opinion of Soviet circles, the rumours of Germany's intentions to tear up the Pact and to undertake an attack on the U.S.S.R. are without foundation. As for the transfer to the northern and eastern areas of Germany of troops during the past weeks, since the completion of their task in the Balkans, such troop movements are, one must suppose, prompted by motives which have no bearing on Soviet-German relations.'

Krushchev was, many years after the War, to cite this Tass communiqué as an example of Stalin's shortsightedness and lack of judgement. Some of Stalin's apologists have described it as a last desperate bid at appeasement. Whatever the motive, Russia's countless millions could be excused for wondering on 22 June whether their leaders knew what they were about. Confidence was badly shaken, and Stalin had to restore morale.

On the anniversary of the October Revolution in 1941, with the Germans standing at the gates of Moscow, he set the theme which he hoped would achieve his purpose, in his speech in the Red Square: 'May you be inspired in this war by the heroic figures of our great ancestors.'

The war became the 'Great Patriotic War'. The appeal was directed specifically at Russian national pride. With the Baltic states and most of the Ukraine gone, it was to the heroes of ancient Muscovy that the people were asked to look for inspiration – to Alexander Nevsky who had routed the Teutonic Knights in 1212, to Dmitri Donskoi who had defeated the Tartars in 1380, to Minin and Pozharky who had fought the Polish invaders. The Soviet régime and Stalin sought to identify themselves with men like Suvorov, who had defeated the Turks and some of Napoleon's armies in Italy, with Kutuzov who had mauled Napoleon's *Grande Armée* at Borodino, and although he could not halt its advance then, had eventually worn it down and driven its shattered remnants from Russia's sacred soil. Patriotism had carried Holy Russia through many a trial and in these times the Orthodox Church and many of its priests had played a distinguished part. Their memory was to be honoured together with those of other heroes of the past. Stalin insisted on cordial relations with the leader of the Orthodox Church. In the interests of national unity, of Russian patriotism nothing was to be done to offend anyone's religious belief. In return the Church became increasingly vocal in its support of the régime, even to the point of saying special prayers for Stalin. Marxist purists may have winced but Stalin was not to be deflected from pursuing his goal: national unity.

Sensing the change of direction in relations between the régime and the Church, Emilian Yaroslavsky, the 'anti-God' leader, had early on in the war devoted an issue of *Bezbozhnik,* the 'godless' magazine, not to denigration of religious teaching and priests but to indignant denuncia-tions of the Nazi persecution of the Protestant and Catholic churches in Germany. This was too blatant a *volte-face* for the Soviet propagandists – because it ran the risk of causing the whole carefully devised theme of national patriotic unity to lose credibility in Russian eyes. *Bezbozhnik* was immediately closed down, and Emilian Yaroslavsky was set to work writing pamphlets on 'The Great Patriotic War'.

Hitler was given the chance of undermining or at least slowing down the momentum of Stalin's 'Patriotic War' theme in the Soviet Union. It came in 1942 when the Soviet Lieutenant-General Vlassov was taken prisoner by the Germans. Vlassov was one of the Red Army's younger commanders who had been promoted rapidly and on the basis of his war record. He had enjoyed a considerable reputation for imaginative leadership and personal bravery among both officers and men, and Stalin was said to have him in mind for higher commands. He had distinguished himself particularly for his part in the successful defence of Moscow the previous winter. He now proposed to his captors to form a Russian Liberation Army from among the millions of Russian prisoners in their hands. Russia, he said, must be cleansed of Bolshevism. He was appalled, he claimed, by millions of Russians being led into battle like sheep to the slaughter. The fault, he maintained, lay entirely with the Bolshevik régime. His aim was to rebuild on the principles of the Kerenski revolution of 1917 – a united and socialistic Russia with autonomy for the non-Russian elements of population.

This last point of course ran directly counter to the long-term plans Hitler had for Russia, and the Führer decided that for the time being Vlassov should be used on the front spreading his message by loudspeaker and short-range broadcast to Soviet troops in order to procure the maximum number of defections.

Vlassov's success in encouraging desertions was impressive – a reflection of the deep dissatisfaction which still prevailed in many sections of the Russian armed forces and population with the Soviet régime.

In 1942, however, Hitler still hoped to settle the Russian question his way. It was only after the disaster of Stalingrad in the summer of 1943 that the project of the Vlassov army was taken up again. Even then the Nazis were extremely reluctant to see the project implemented, and it was not until April 1945 – a few weeks before the end of the war – that one Vlassov division was eventually activated. It fought in one action and acquitted itself well. After the armistice Vlassov was handed over to the Russians. He was hanged in 1946, together with several of his fellow officers.

Although Hitler never intended the project of General Vlassov, a Great Russian by ethnic origin, to be implemented, the Wehrmacht commanders did succeed in impressing on him their ever-increasing shortage of manpower, and under pressure from them he grudgingly gave his consent in 1942 and again in 1943 to the recruitment of volunteers among the national minorities living within the borders of the Soviet Union.

As it turned out his agreement came too late. Admittedly, a few volunteer units were raised in the Baltic countries and fought with the Germans; and several larger units were recruited in the Ukraine. They were for the most part used not for fighting against the Russians but as occupation troops in Western Europe, particularly France, to release German forces for fighting in Russia and on other fronts. As for the rest, the overwhelming majority who had cheered the Germans as liberators in 1941 and had begged to be allowed to fight at their side against the Soviet oppressors, was no longer interested in volunteering for Germany in 1942, and still less so in 1943. Germany's repressive measures – the rounding-up of men and women for forced labour, near starvation rations, innumerable executions, often in public, at the slightest or even without provocation – had turned yesteryear's potential friends into embittered enemies, and many of those who were fit and strong enough, had taken to the woods and marshes to join partisan bands. To begin with, these partisans were frequently as anti-Soviet as they were anti-German. The same was true of many partisans in Great Russia. But Stalin's theme of the 'Great Patriotic War' eventually offered them – or at least most of them – a way back into Soviet society.

As far as Hitler was concerned, however, his belated agreement clearly demonstrated that he had missed the bus – not that he himself was probably aware that there had been a bus to miss.

In contrast to Stalin, Hitler had no moral problems to face as a result of the outbreak of the Nazi-Soviet war in 1941 either in the Wehrmacht or among the civilian population at home. Like Stalin, he had to keep quiet about his true feelings about the other side's ideology between 1939 and 1941, but in the light of his numerous vitriolic speeches against Bolshevism neither Nazis nor non-Nazis in Germany doubted that he would attack the Soviet Union sooner or later. They accepted it as inevitable, and when the moment came there was little enthusiasm, but there was also a quiet confidence in victory.

While that confidence remained unshaken, Moscow's message had no hope of making an impact. The Soviet Radio stressed that this was a war not between peoples but against Fascist criminals. There were 'two Germanies'; the fight was against Hitler and his minions whom the Germans had allowed to gain control over their nation; now was the time for the Germans to throw off Hitler and his criminal régime. It was all to no avail. Bolshevism and Stalin were distrusted too deeply.

Would that distrust diminish once the tide of war turned after Stalingrad? Certainly the number of Germany's listeners increased, by all accounts. And for a very simple reason: many Germans hoped that one of the German language transmissions from Russia might perhaps give them some clue as to what had happened to friends or relations who had been posted missing on the Eastern Front. After Stalingrad, with Germany's armies in retreat, the Wehrmacht found it impossible to keep accurate records of who had been killed in action, wounded or taken prisoner, and, in case of doubt, it had no alternative but to post a man missing. From early 1943 onwards an ever-increasing number of

German families received this notification, and their only hope of learning more about husbands, fathers, brothers and sons lay with Radio Moscow.

The Russians adopted the practice, when they wished to give news of German soldiers, of directly calling their next-of-kin to the radio set. On the pretext of allowing time for a neighbour to fetch this next-of-kin – in case he or she was not listening already – Radio Moscow would then broadcast a political talk or an analysis of the political situation. Eventually, the announcer would address himself personally to the next-of-kin: 'Herr X of Leipzig, your son was captured in the fighting north of Kharkov. He was badly wounded but our doctors are looking after him, and he will be all right. You are very fortunate Herr X. Your son will come back to you after Hitler has been defeated. Millions of German fathers and mothers are not so fortunate. Their sons will not come back. Hitler's criminal ambitions have cost them their lives.' Or: 'Frau Y of Magdeburg, your husband died on the Berezina. Frau Y, what was your husband doing deep inside Russia on the Berezina?'

The theme in all Russian-controlled broadcasts was that the German people had been misled, that Hitler and his Nazi gang were the real criminals, that Russia's quarrel was with him and not with the German people. There is little or no evidence that these broadcasts evoked any positive response.

In July 1943 the Russians adopted an entirely new line in trying to win support in Germany. They had noted that Communism held little appeal for the majority of Germans. Even in the German prisoner-of-war camps in Russia German Communists who had fled to the Soviet Union when Hitler seized power in Germany, men like Walter Ulbricht and Wilhelm Pieck, had made few converts to Marxism. Was it therefore not time to try a new approach? After all, Stalin had steadied Russian morale by his appeals to patriotism, to nationalism, to the glorious past. Why not try to shake Hitler and the Nazis with appeals to German nationalism?

July 1943 saw a National Committee of Free Germans set up. Its president was the well-known German Communist writer, Erich Weinert, and vice-presidents Major Carl Hetz and Lieutenant Count von Einsiedel, a great-grandson of Bismarck's who said that it had always been his grandfather's golden rule not to attack Russia and the West at the same time. Communists like Ulbricht and Pieck were, of course, members of the Committee, but they kept discreetly in the background. The accent was on German nationalism, and the flag at the inaugural meeting bore the black, white and red colours of pre-

1918 Germany which had also been adopted by many right-wing political factions in the days of the Weimar Republic. The manifesto issued by the meeting contained passages such as:

'If the German people continue inertly to follow Hitler, then he can be overthrown only by the armies of the Coalition. But that would mean the end of our national independence and the partition of our country.'

'If the German people have the courage to free Germany of Hitler ... then Germany will have won the right to decide her own fate, and other nations will respect her. . . . But no-one will make peace with Hitler; therefore the formation of a genuine National Government is an urgent task.'

'The forces in the Army, true to their Fatherland, must play a decisive part in this. Our aim is a Free Germany, i.e. a strong democratic power totally unlike the impotent Weimar Republic. . . .'

In September 1943, two months later, a League of German Officers was formed with General von Seydlitz, a descendant of Frederick the Great of Prussia's famous cavalry commander at its head. In fact, the membership of the League reads like a list of Germany's aristocratic military families.

The League and the Committee were integrated. They had put at their disposal a 'Free Germany Radio Station' to spread their message into Germany. They also produced a paper called *Freies Deutschland,* hundreds of thousands of copies of which were showered on the German lines, as well as millions of leaflets. One of these showed a Valkyrie-like Germania who had broken her shackles, driving a huge sword into a spider-like Hitler. The caption was the Nazi slogan *'Deutschland Erwache'* – Germany awake. Another leaflet showed Count von Einsiedel in his Luftwaffe Uniform, holding up an awe-inspiring picture of Bismarck who was pointing to a scruffy Hitler and saying: 'This man is leading Germany towards catastrophe.' Count von Einsiedel is reported as warning: 'A war of Germany against Russia is stupid and without a chance of success. This my great-grandfather, Otto von Bismarck, said over and over again, and every soldier must daily become more convinced of its truth.'

Another important function of General von Seydlitz and members of the League of officers was to use loudspeakers to persuade encircled German formations to surrender. When ten German divisions were surrounded at Korsun, General von Seydlitz claimed to 'have saved 20,000 men'.

As a rule, the Russians soon learned that it was easier for generals to surrender than for ordinary soldiers. During the Soviet summer offensive of 1944 no fewer than 22 generals surrendered in a five-week period. The reason for this is to be found in practical realities involved in surrender. Generals as a rule have wireless equipment at their disposal with which to arrange the necessary details carefully in advance. Ordinary soldiers enjoyed no such facilities, and it could be highly dangerous to cross the line. Someone from your own side might shoot you in the back as you are trying to make your get-away, or somebody on the other side may misunderstand your action and shoot first without bothering to ask questions.

Moreover – more important – the ordinary German soldier knew of the brutalities – the mass transportation of slave labour, the public executions, the starvation and the looting – that had been carried out by the invaders in Russia. He knew the ordinary Russian had no reason to like the Germans. What if the Russian to whom he surrendered had lost his wife, his brother, or his parents had been slaughtered? Could he expect any mercy when so many Germans had shown so little? The ordinary German soldier preferred to fight his way out of encirclement across mountain passes and rivers against numerical odds, just so long as there was a chance of moving a little closer towards Germany rather than face the grim uncertainties of surrender. Hitler's deliberate policy of brutality in Russia militated against the surrender of the rank and file of the German army.

Early in 1945 the League of German Officers and the National Committee of Free Germans was quietly phased out.

Hitler's arrogant and brutal contempt for the 'sub-human' Slav contributed to his losing the Battle for Russia; it also foiled Russia's attempt to bring the fighting on the Eastern Front to an end before Soviet tanks had to crash their way into the Wilhelmstrasse in Berlin.

CHAPTER 11

The U.S.A.:
Propaganda on Two
Fronts

The United States entered the First World War by a deliberate act, by formally declaring war on the Central Powers. The country at large reacted to this act not with demonstrations of enthusiasm or jubilation but with fatalistic acceptance, and this mood continued for several months until President Wilson, with the aid of George Creel, roused the nation by turning the War into a crusade for freedom, liberty, self-determination, humanity and, in H. G. Wells' phrase, called it 'the war to end war'.

In the Second World War the United States was bombed into becoming a belligerent by the Japanese attack on Pearl Harbor. Before Pearl Harbor most Americans were pro-Ally, but most of them seemed equally determined not to get actively involved themselves, and nothing, so it appeared, could shake this 'pro-Ally, anti-War' mood. After Pearl Harbor even the most vehement anti-interventionists, with few exceptions, fully and wholeheartedly supported the American war effort. There was no choice but to fight and win. 'Remember Pearl Harbor', was a rallying cry that united the nation.

Yet more is needed than a rallying cry to keep a nation and its allies united in the long, hard grind of a World War. President Roosevelt who, like Churchill in Britain, was his nation's most effective propagandist at home and abroad, had spelled out in general, broad terms, the ideals America was seeking for herself and the rest of the world, long before the United States became a belligerent. He had committed his country to the Four Freedoms – Freedom of Speech, Freedom to Worship God for everyone in any part of the world in his or her own way, Freedom from Want, and Freedom from Fear – and he had elaborated on these Four Freedoms in the Atlantic Charter which he and Churchill signed in Argentia Harbour, Newfoundland, in August 1941.

Long-term, strategic aims, however, are not enough as the fortunes of war change, often in unpredictable ways. They certainly had little psychological impact either in America or among America's allies as Japan followed up her attack on Pearl Harbor with a seemingly interminable succession of victories in the Pacific and South East Asia. Indeed, it seemed that she could not be stopped wherever and whenever she chose to attack. The world was as horrifyingly mesmerized by Japan's apparent invincibility in 1941 and early 1942 as it had been by Hitler's in 1940.

The result in America was that a great many people urged that their country should conserve its manpower and material resources for its own defence instead of sending men and supplies to Europe, to the Pacific, to Russia and to Chiang Kai-shek's China. It was a trend of opinion that gathered so much strength that Roosevelt, the propagandist-in-chief, decided to intervene personally. He announced that he would broadcast one of his famous 'fire-side chats' to the nation on the occasion of George Washington's birthday in February 1942. He gave notice in advance that the purpose of his broadcast was to explain his world-wide strategy, and listeners were advised to have maps ready in order to be able to follow his arguments and to understand why a 'Defend America only' strategy was defeatist.

President Roosevelt began by saying that this was a new kind of war. 'It is different', he continued, 'from all other wars in the past, not only in its methods and weapons but also in its geography. It is warfare in terms of every Continent, every island, every sea, every air lane in the world. . . . There are those who still think in terms of sailing ships. They advise us to pull our warships and our planes and our merchant ships into our home waters and concentrate solely on last-ditch defence. Look at your map. . . . It is obvious what would happen if all these great reservoirs of power (Britain, Russia, China, Australia, New Zealand) were cut off from each other by enemy action or by self-imposed isolation. . . . From Berlin, Rome and Tokyo we have been described as a nation of weaklings – "playboys" – who would hire British soldiers, or Russian soldiers, or Chinese soldiers to do our fighting for us.'

Roosevelt paused. And when he started talking again, it was more slowly, more deliberately to achieve the maximum dramatic impact. It is small wonder that his 'fire-side chats' were carefully analysed in Goebbels' propaganda institutes as ideal examples of effective broadcasting techniques.

'Let them repeat that now', he went on.
'Let them tell that to General MacArthur and his men.'

German
Christmas
IN
JUGOSLAVIA

13. German script for 'Christmas' used to emphasize the contrast between Teutonic sentimentality over 'Weihnachten' and Nazi brutality in occupied Europe.

AVE CAESAR!

Morituri te salutant

14. Italian enthusiasm for the war, never strong, was not encouraged by the sight of their Duce – Hitler's puppet – despatching their soldiers to death.

'Let them tell that to the sailors who today are hitting hard in the far waters of the Pacific.'

'Let them tell that to the boys in the Flying Fortresses.'

'Let them' – and he paused for a long time – 'tell that to the Marines.'

While Roosevelt was speaking, a Japanese submarine surfaced off the coast of Santa Barbara, California, and fired a few rounds into a nearby ranch. The shelling caused no casualties and only minor damage but it produced huge headlines right across the country and practically squeezed Roosevelt's 'fire-side chat' off the front pages.

On the face of it, the Japanese exploit should have served to destroy the argument Roosevelt was trying to get across to the American public in his broadcast and reinforced the case of those who wanted to see a 'Defend America only' strategy pursued. Fortunately, the Japanese were so unpopular at the time and – more important – were regarded as so devious and treacherous that any move by them intended to make the Americans concentrate their forces on defending their homeland was immediately suspect as another trick or trap to be avoided at all costs. Whatever the Japanese expected one to do, it was best to do exactly the opposite. So, perversely, the Japanese submarine's attack on the farm near Santa Barbara helped to strengthen Roosevelt's argument instead of weakening it.

As for Roosevelt and his propaganda experts, they absorbed the lesson that it is not always wise to announce important broadcasts and leak hints of their likely contents too far in advance.

Japan's spectacular successes in 1941 affected opinion not only in America but also on the other side of the Atlantic. Goebbels was jubilant. Who, he asked condescendingly, could still seriously contemplate America as the powerhouse of the so-called 'United Nations' war effort? The poor British and the Russians could expect no tanks, no planes, no munitions from the United States. What munitions the United States had were being lost in the Pacific at a faster rate than American industry could replace. All American soldiers, sailors or airmen were absorbed in the Pacific war; there were none to spare for Europe. And even if Roosevelt could scrape together a battalion or two or a few airforce squadrons to send to Europe, they were bound to be too late to make any difference to the outcome of the War; by the time they arrived, the Germans and the Japanese would have linked up in Asia. The issue would have been settled.

Many in occupied Europe who longed to be freed from German occupation reluctantly shared Goebbels' view. Indeed, morale there had never sunk lower since the fall of France in 1940. Roosevelt's long-term strategic aims for the future of the world were all very well, but these aims could be realized only if the United Nations beat Germany, Italy and Japan and won the War. What America had to disseminate to friend and foe alike at this period of the War, was less a set of lofty principles but more emphatically a picture of America's power and capacity to wage war successfully on two fronts across two mighty oceans. Only thus could it hope to undermine the morale of the enemy and restore confidence amongst its allies and the peoples of occupied Europe.

Before Pearl Harbor, President Roosevelt had set up the Office of the Coordinator of Inter-American Affairs (CIAA) under Nelson Rockefeller, and the Office of the Coordinator of Information (COI) under Col. (later General) William J. Donovan, which included a Foreign Information Service (FIS) under Robert Sherwood. The FIS had actually started shortwave broadcasts to Europe several months before the U.S. entered the War and began short-wave broadcasts to Japan from the West Coast within hours of the Japanese attack on Pearl Harbor. Nelson Rockefeller's CIAA had, it was generally agreed, been successful in convincing most of Latin America not to allow themselves to become involved with the Axis powers – no easy task since many Latin American countries had large numbers of settlers from Germany and Italy who frequently held key posts in politics, industry, banking and the armed forces. The CIAA could even point to some positive gains in its campaign to persuade Latin Americans that their interests were best served by the policies pursued by the United States. When it came to reorganizing the propaganda machine, it was, therefore, decided to leave the CIAA untouched.

The rest of the machinery, however, was plainly inadequate, for the job required to sustain America's propaganda effort on the strategic and tactical level and on two fronts.

In June 1941 the COI under William J. Donovan was disbanded, and two new agencies were created, the Office of War Information (OWI) and the Office of Strategic Services (OSS). The OWI was divided into a Domestic Branch and an Overseas Branch which was in fact built up around the former Foreign Information Service and retained Robert Sherwood as its chief.

As Director of the OWI Roosevelt picked the well-known journalist and broadcaster Elmer Davis because he felt Davis had a keen yet humane understanding not only of his own countrymen but also of people in other countries. He trusted Elmer Davis' judgement of people and loved to recall the comment Elmer Davis made on the 'fire-side chat' on George Washington's birthday in February 1942 : 'Some people want the United States to win so long as England loses. Some people want the United States to win so long as Russia loses. And some people want the United States to win so long as Roosevelt loses.'

It was a comment which turned out to be prophetic – certainly in so far as the last sentence was concerned. Elmer Davis and a succession of harried directors of the Domestic Branch of OWI found themselves in almost constant trouble with Congress. When OWI brought out an anti-inflation tract and a heroic cartoon history of the President and Commander-in-Chief, the House of Representatives was so furious that it abolished the Domestic Branch outright and only grudgingly allowed the Senate to save it. Of course George Creel in the First World War had to cope with similar problems. In the Second World War as in the First, Congress was suspicious of anything that smacked of internal party-political propaganda, and much of OWI's time was taken up – as George Creel's had been – in justifying its work and explaining that public money was not being spent in promoting any political personality or taking sides in any domestic political issue.

The other new agency, the Office of Strategic Services (OSS) which was placed under William J. Donovan, was also given psychological warfare functions, specifically in connection with military operations, and was made responsible to the Joint Chiefs of Staffs.

Inevitably there were frictions between the two agencies, and although Roosevelt re-defined their respective functions and areas of competence by Executive Order, his re-definition was open to different interpretations, and so from time to time clashes between the two agencies continued to occur right up to the end of the War.

To add to the confusion, both the Navy and the Army took an interest, though by no means a consistent interest, in psychological warfare. Furthermore, in the vast commands that grew up during the Second World War in the Pacific, the Atlantic and in Africa and later in Europe, it was left to the discretion of each theatre commander to decide what, if any, form of propaganda or psychological warfare he wanted carried out in his area. Some, like Admiral Halsey, refused point-blank to give OWI personnel clearance for his area of command. He, like a good many other senior American officers in the Pacific, were convinced that it was a waste of time and effort to try to undermine the morale and perhaps even induce the surrender of Japanese soldiers who had been taught that they ceased to be members of Japanese society the moment they were taken prisoner, who fought tenaciously and ruthlessly even when outnumbered and outgunned and who, when trapped deep in some cave on a remote Pacific island, blew themselves up with a grenade rather than be taken alive.

Not all American commanders shared this view. General MacArthur encouraged psychological warfare in his theatre of war, and Admiral Nimitz gradually came round to the view that psychological warfare could be a help in military operations.

The structure of the American propaganda organization can hardly be called tidy. Yet despite the untidiness, the division of authority, the clashes between different agencies, government departments and commands, the OWI gave a unity to America's propaganda effort. It made a close study of the enemy target areas – the Axis powers in Europe and Japan in the Far East. Under Elmer Davis' gentle, but firm direction the OWI in its broadcasts laid the emphasis on facts rather than polemics. Not that he left his listeners in any doubt about what Americans felt about Nazism or the militarist régime in Tokyo, but he impressed on his staff that what was likely to have the greatest impact on even the most dedicated Nazi or supporter of General Tojo was news of American resilience, of her growing power which her enemies could never hope to match and which was bound to outstrip theirs.

The best way to deflate Goebbels' claims that America's resources were being inexorably sucked into the Pacific war against the Japanese and that what resources the United States had to spare were being sunk by Nazi U-boats off the North American coast at an alarming rate – until the U.S. imposed a partial black-out on its Atlantic coast towns and villages in April 1942 – was to point out that the number of American divisions in Britain was increasing week by week. Were the Germans aware that the American ground forces in Britain were large enough by June 1942 – just over six months after Pearl Harbor – for an American commander-in-chief to be sent to London, one General Dwight D. Eisenhower? Were they also aware of the fact that Londoners – irreverent as always – had renamed Grosvenor Square in the heart of London's Mayfair, 'Eisenhowerplatz' because so many American generals, admirals and staff officers were constantly arriving for or leaving from conferences? Furthermore, had Geman soldiers stationed in Western Europe written to their families at home yet about the new heavy bombers that were appearing in the skies with up to then unfamiliar markings on their wings and tails? Well, the markings were American, and the bombers were Flying Fortresses. Half a year after Pearl Harbor, the U.S. Eighth Air Force had arrived in Britain.

Elmer Davis insisted on a similar approach in broadcasts directed towards Japan. To attack the Emperor, his instructions laid down, might prove counter-productive. Admittedly, he presided formally over the Supreme War Council, but he was also regarded as semi-divine and to insult a god-emperor might stiffen resistance instead of weakening it. All OWI personnel had to remember was that America's quarrel and that of her Allies was with Japan's militarist régime and not with Japanese monarchy; the future of the Japanese monarchy with its religious connotations was a question for the Japanese to decide on in due course. In the meantime, the target to attack was the militarist régime. And nothing was likely to impress the members of that régime and those who for the time being had to go along with it – be they members of the imperial family or ordinary citizens – more than America's growing power.

If the Japanese had destroyed the U.S. Pacific Fleet at Pearl Harbor in December 1941, how was it that Admiral Nimitz could face Admiral Yamamoto at Midway in June 1942? Did the Japanese know that in the Battle of Midway Yamamoto lost all of his four fast aircraft carriers, leaving the Japanese Navy no more than five carriers, of which only one was a large one? Six more carriers were being repaired or built in Japanese naval yards, but did the Japanese know that nearly three dozen aircraft carriers were nearing completion in American naval yards and that all of them were destined for use in the Pacific? Moreover, American war production was only just getting into its stride as indeed was the mobilization of American manpower. If the Japanese Imperial Army was so invincible, why was it not already in occupation of Australia instead of battling for every inch of ground in the Solomon Islands? It had no hope of reaching Australia, and as American power expanded, the Japanese would find themselves overwhelmed by America's manpower and material resources. The Japanese people had allowed themselves to be misled about the realities of the situation; the militarist régime which ruled the country was infected with the 'Victory Disease', and the 'Victory Disease' was leading Japan to disaster.

Elmer Davis' instructions on the way broadcasts to Japan were to be conducted were explicit: broadcasters must sound calm and polite at all times, never strident; they must observe all the elaborate courtesies that the Japanese language and Japanese customs call for; to shout and scream – even to raise one's voice – will only antagonize listeners and cause them to switch off, quite apart from the practical detail that it is likely to make short-wave listening technically impossible.

OWI personnel were reminded that however harsh the sufferings some members of their own families in the war against Japan might be, they were talking to people who were themselves suffering great hardships and who had to be convinced that their lot could be eased only if the present militarist régime released its hold on their country. The point to impress on the Japanese was that their country could not win; and the most effective way of doing so was, first, to be as factual as possible and, secondly, to keep the news as simple as possible. Under no circumstances were too many names to be used. The Japanese could not be expected to attach much meaning to the dozens of American and British generals and admirals who were holding important command posts. So OWI personnel was instructed to be highly selective. Persons like Roosevelt, Churchill or MacArthur meant something to them. Few of the other Allied leaders did. The same held true of place names. The geography of the Pacific was difficult enough; the geography of North Africa and Europe was virtually unknown. It would have been meaningless to a Japanese audience to enumerate all the places that Montgomery's Eighth Army recaptured after defeating Rommel at El Alamein. The important point to impress on a Japanese audience was that the Germans were not turning out to be quite as powerful military allies as Tokyo's militarist régime had suggested; therefore the gist of any news item on El Alamein had to be that the Germans were about to be thrown out of Africa. By the same token, the important aspects of the German *débâcle* at Stalingrad was not the name of the town but that it was deep in Russia, not the name of the German commander, Paulus, but the fact that he was a field-marshal, not the number of the German army which surrendered to the Russians, the Sixth, but the number of men in that Army – over 300,000 men.

In other words, the OWI developed a consistent, coherent philosophy and approach to propaganda both against the Axis powers in Europe and against Japan. And despite the untidy structure of America's propaganda machinery – in fact perhaps because of it – this philosophy and approach made itself felt throughout America's diffuse propaganda outlets and acted as a sort of unifying element. For whenever a special psychological warfare operation was set up, the OWI was usually consulted and, as often as not, was asked to detach men to such an operation. Naturally these men brought with them the thinking, the training, the methods, they had learned in the OWI.

Elmer Davis' and the OWI's main problem was lack of contact with the top policy makers. Unlike his British opposite number, Sir Robert Bruce Lockhart, the head of the Political Warfare Executive, who was not only a close personal friend of Anthony Eden, the Foreign Secretary, but was also kept informed on all important policy questions by a ministerial committee of three, presided over by Eden, Elmer Davis was rarely, if ever, in the know. He was on personally cordial terms with President Roosevelt and on paper had the right of direct access to him, but in practice he found – unlike George Creel in the First World War who frequently saw President Wilson several times a day – that it was difficult to see the President and that on the rare occasions that he did, that there were others there whose presence made discussion on matters of high policy impossible. Elmer Davis and the OWI therefore largely operated in the dark.

Working under such a handicap, an accident was bound to happen sooner or later. It did late in July 1943 just after King Victor Emmanuel had dismissed Mussolini and appointed Marshal Badoglio Prime Minister. The OWI put out a broadcast quoting a commentator who referred to Victor Emmanuel as that 'moronic little king' and to Badoglio as 'the butcher of Abyssinia'. The United States and Britain were at that moment involved in highly secret and delicate negotiations with Italy to arrange a separate armistice. Roosevelt was appalled at the broadcast and tried to save the situation at a press conference. 'When a victorious army goes into a country,' he said, 'there are two essential conditions that they want to meet. The first is the end of armed opposition. The second is when that armed opposition comes to an end to avoid anarchy. In a country that gets into a state of anarchy, it is a pretty difficult thing to deal with because it would take an awful lot of troops. I don't care with whom we deal in Italy, so long as it isn't a definite member of the Fascist Government, as long as they get them to lay down their arms, and so long as we don't have anarchy. Now he may be a King, or a present Prime Minister, or a Mayor of a town or a village. . . . You will also remember that in the Atlantic Charter, something was said about self-determination. That is a long-range thing. You can't get self-determination in the first weeks that they lay down their arms. In other words, common sense. . . .'

Roosevelt's press conference went some way towards putting right the impression produced by the OWI broadcast, but all the trouble could have been avoided if he or someone at the State Department had informed Elmer Davis of the secret armistice negotiations with Italy.

Curiously enough, it was the joint Anglo-American Psychological Warfare Branch at General Eisenhower's Headquarters in the Mediterranean – a Branch which of course included members of the OWI – that several weeks later saved the armistice agreement, and Italy's decision to change sides, from becoming unstuck at the last minute. It had been agreed that Marshal Badoglio should go on the air on Rome Radio at 18.30 hours to announce the armistice and Italy's change of sides. The timing of the Marshal's broadcast was vital because on the following morning the Allies were to land at Salerno. Unless the Marshal himself broadcast the announcement, thereby validating the armistice as authentic and genuine, the Allies would have had to face not only the German divisions which had been pouring into Italy but the whole Italian army.

A few hours before Badoglio was due to go on the air, a message reached Allied Headquarters from the Italian Government that the whole armistice was off. The Psychological Warfare Branch was asked if it could think of a way out. Various suggestions were brought up and discarded. Radio Algiers should simply broadcast the text of the armistice agreement but this, on further consideration, was thought liable to be treated with incredulity and merely as a propaganda trick. Another idea was to broadcast the agreement simulating Badoglio's voice, but this was felt likely to cause nothing but chaos and confusion, especially if a denial was broadcast from Rome immediately afterwards. Of course no one knew what was happening in Rome, what had caused Badoglio to change his mind. The fact that he had managed to get a message through suggested that he was still a free man and not a prisoner of the Germans. If he was free, was there then any way to make him change his mind again? The members of the Psychological Warfare Branch pooled what facts they knew about the Marshal. He had always held himself slightly aloof from the Fascist régime. He liked to be thought of as a professional soldier, a man of honour, an officer and a gentleman. He was certainly not someone who would like to be seen and known to have broken his word. After all, many senior Italian officers and members of his staff had been involved in the armistice negotiations. Here, in the view of the Psychological Warfare Branch, lay a possibility of a way out. There was no time for an exchange of messages. All had to be staked on one broadcast.

At 18.30 hours exactly Radio Algiers went on the air with a message from General Eisenhower. The announcer then explained that it had been foreseen that the Germans might prevent Marshal Badoglio from announcing the armistice from Rome and that it had been agreed between the Allies and the Italians that if that happened, the Marshal's message should be read over Radio Algiers – which the announcer then proceeded to do.

An hour later Marshal Badoglio was on the air from Rome.

The most surprising and fascinating successes by psychological warfare teams, composed largely of OWI personnel, were attained in the Pacific and the Far East. There – at least in the theatres of war where they were allowed to operate – they achieved what many had thought impossible: to make Japanese soldiers surrender and co-operate with their captors. The key to their success was careful analysis of the way

Japanese soldiers behaved after being taken prisoner. It was noted, for example, that a great many Japanese soldiers claimed that they had been asleep or unconscious when they were captured, claims which often clashed with the accounts of the American soldiers who reported that the Japanese put up a stiff fight at first, but gave up when they must have regarded the odds as hopeless. As several American Marines, when told of the Japanese explanations, remarked somewhat wryly, 'for soldiers given to fainting fits they are pretty tough fighters'. The interesting aspect, as far as the psychological warfare teams were concerned, was that not all Japanese soldiers were prepared to fight to the death, but that since they knew that as prisoners they were according to the Japanese army code no longer members of Japanese society, they thought up some excuse like being asleep or unconscious, which they hoped would later allow them back into Japanese society.

Even more fascinating was the discovery that some Japanese, on being captured, immediately volunteered to fight on the side of the Americans and appeared genuinely downcast when their offers were rejected. Denied the opportunity of fighting with the Americans, a few Japanese then pleaded to be taken up in a 'plane to point out Japanese gun emplacements and the areas where there were the largest Japanese troop concentrations. To begin with, the Americans found these almost instant changes of loyalty and allegiance incomprehensible, but whenever the local American commander was prepared to take the chance and authorized a Japanese prisoner to be taken up to point out Japanese positions, the information always turned out to be accurate.

Sifting through the available evidence the psychological warfare teams came to the conclusion that the Japanese army code was directly responsible for the Japanese behaviour. By insisting that soldiers who were taken prisoner ceased automatically to be members of Japanese society, the code prevented any instruction to be given to soldiers on how to behave when taken prisoner. Japanese were not told that they had to give only their name, rank and service number, and nothing else. They were also not taught that, if possible, it was their duty to escape. The code made them outcasts and deprived them of all the signposts by which their lives had been regulated previously. Being outcasts of one society, they naturally tried to find a place in another. Hence the curious phenomenon of the almost instant change of loyalty on being captured and the offers of fighting with the Americans.

Having traced the reasons for the psychological state of Japanese prisoners of war to the Japanese army code, the OWI teams sought to extract the greatest possible advantage out of the situation.

One objective naturally was to obtain information about the target area at which the OWI, its various branches and other agencies were directing their propaganda. It is here that the work of the OWI was of particular significance because, with the exception of a few people who knew Japan and the Japanese, the picture the vast majority of Americans or, for that matter, her allies, had of the Japanese was strictly stereotyped, and stereotyped pictures of a propaganda target tend to be based on prejudices and generalizations which bear little relation to the truth. Consequently, propaganda can in such circumstances have little effect. It is bound to be wide of 'the target'.

The work of the psychological warfare teams helped to build up a true picture of the target. The information Japanese prisoners of war provided, was of course, in the first instance, military intelligence in the war area in which they were captured, and then valuable intelligence about the Japanese armed forces as a whole. At this stage intelligence and more general information useful for propaganda purposes began to overlap. For instance, morale was a subject which was central to the OWI teams' purpose. It transpired that militarist indoctrination had seeped right through the Japanese war machine, that there were some units in the Japanese armed forces which were more fanatical in their determination to fight to the death than others. Prisoners of war also threw a penetrating light on home morale. Without regarding themselves as in any way disloyal to their Emperor or Japan's traditions, it appeared that there were many Japanese who harboured serious doubts about the philosophy and wisdom of the policies pursued by the ruling coalition of aggressive militarists and financiers.

One of the most useful sources of information was a discussion group an OWI psychological warfare team in Burma formed of Japanese prisoners of war. All the members of the group were given false identities for their own protection. They included officers and other ranks, men who had been to university and workers. What they shared was a wish to see a different kind of Japan after the war, a Japan which retained what was best in its traditions but discarded what was worst. They spent a good deal of their time talking about the future, but they were agreed that nothing could be done about the future until the war which they felt Japan was bound to lose, was over. And so after much discussion among themselves, they agreed to help shorten the war into which an irresponsible militarist clique had led their country.

It was on their advice that the language used in the Surrender Passes dropped on Japanese lines was formulated. The surrender appeal was in the dignified military terms a senior officer, with the best interests of his country and his troops at heart, would use. This type of Surrender Pass proved highly effective.

In some war areas the psychological warfare teams used Japanese prisoners of war to broadcast appeals to surrender by loudspeaker. The extent of success varied. In the Marianas one junior Japanese officer was extremely effective to begin with. His message was very direct and simple: Since Japan was everywhere on the defensive and it was a question not of winning or losing but of how soon Japan would lose, the sooner the war was brought to an end the better; otherwise the destruction which Japan would suffer at home and the number of killed and maimed would be tremendous; to minimize the suffering of Japan, his compatriots should surrender; they would not lose face by so doing; the only ones who would lose face were the Americans who under International Red Cross rules were compelled to provide them with food, clothing and shelter; they were well provided for, and the more Japanese surrendered, the greater the strain would be on America's resources. So surrender, increase the strain on America's resources and at the same time save Japan from further destruction.

All this was delivered in formal military language, and the response was encouraging. But after a week the number of Japanese who surrendered tapered off to zero. A Japanese-speaking member of the psychological warfare team went out with the Japanese junior officer on his next loudspeaker mission to the front. What he heard, appalled him. Gone was the formal military language of the early broadcast. Gone, too, was the simple directness of his appeal to surrender. He now talked about the benefits of democracy, the joys of free speech and the glories of the American Constitution. And all this was delivered in the easy-going manner of an American Forces Network disc-jockey. Plainly the junior Japanese officer had lost his usefulness in that particular task.

Naturally in a field like propaganda which requires imagination, imagination occasionally runs away with its practitioners. This happened during the battle for Okinawa in April 1945. A Japanese lieutenant surrendered with the Okinawa nurse with whom he had been living. He explained that Japan was losing the war, that he wanted to save his country from further destruction and that he would co-operate with Americans in any way that was helpful. He made only one condition: that he be allowed to continue living with the Okinawan nurse who had surrendered with him.

The local psychological warfare advisers were all for accepting his condition. Of course they could not just be allowed to live together; there had to be a proper wedding ceremony. The widest publicity should be given to the occasion, and pictures of the ceremony dropped on the Japanese lines to demonstrate how humane the Americans were.

The couple were in due course married by an U.S. Army chaplain in a Christian ceremony – although both the Japanese lieutenant and the Okinawan nurse were Buddhists – lots of photographs were taken of the ceremony, and the newly-weds spent their wedding night in a tent provided by the U.S. Army.

The photographs of the ceremoney evoked no response from the Japanese soldiers on Okinawa and, once it became known in the United States, nothing but bitter criticism and ridicule.

A much subtler, better planned and executed operation was the broadcasts by Captain (later Admiral) Ellis Zacharias to Japan. It was carried out by the highly secretive Special Warfare Branch (OP–16W) within the Navy Department's Office of Naval Intelligence in close co-operation with the OWI. In concept and philosophy the views of the OWI and OP–16W on this operation were identical, and the two organizations therefore worked together smoothly and effectively.

The idea was conceived about the middle of 1944 when all available reliable intelligence material indicated that some members of Japan's political élite with access to the Emperor were looking for a way of ending a war which Japan appeared bound to lose, but without, by surrendering, impairing the status of the Emperor and what was felt to be 'the national structure of Japan'. The most important members of what for convenience sake was described as 'the peace faction', were identified as Japan's admirals. Captain Zacharias and his colleagues knew that there was no point in appealing to 'hard-liners' such as General Tojo and General Koiso who had nothing to lose by continuing Japanese resistance to the bitter end and who – however mistakenly – believed sincerely that Japan was in honour-bound to do so.

There was also little point in addressing oneself to the Japanese people. Many of them could and did of course hear Captain Zacharias' broadcasts, but no one concerned with psychological warfare seriously entertained even the vague possibility that Japanese public opinion could at that stage be effective. It was an élite, consisting perhaps, of some 500 people or so – generals, admirals, industrialists, political and financial leaders – who had the power to decide whether the nation should fight on, even once its home-islands were invaded, or whether it should surrender. The argument was within that élite, and it was to the so-called 'peace faction' within that group that Captain Zacharias was to address himself.

The 'target' was thus most meticulously defined.

The next issue to be decided upon was 'the message'. According to intelligence the most important element within the 'peace faction' was the admirals. By mid-1944 the Imperial Japanese Navy had ceased to be an instrument capable of offensive naval warfare. It could at best no longer seize the initiative, it could certainly not follow through on any advantages gained by its initiatives, and because of America's bombing strategy of the Japanese home-islands, losses in ships and skilled sailors could no longer be made good by Japanese shipyards or adequate training of fresh recruits. At worst the Japanese Imperial Navy could, by skilful and careful use of its remaining resources, hope merely to delay or spoil U.S. naval attacks in different quarters of the Pacific, knowing that the U.S. naval, Marine and army forces would return to the attack where they so planned at a time of their own choosing – which could mean days or a few weeks.

OP-16W and its OWI advisers decided that it would not only be pointless but counter-productive to drive this point home in Captain Zacharias' broadcasts. The 'peace faction', largely composed of admirals, knew the true position of the Imperial Japanese Navy already. If they did not, they would hardly have been members of the 'peace faction', and if the U.S. made too much play of what must have been plain to them, then U.S. psychological warfare was not merely antagonizing potential supporters within Japan for ending the war as soon as possible, worse still there was a risk of driving the admirals into the arms of 'hard-liners' like General Tojo who were still a powerful element within the Japanese power élite.

It was accordingly agreed what was not to go into the 'message', namely, that the Japanese Navy was not to be made fun of or have its nose rubbed in the mud. No 'message', however, can be entirely negative – unless its purpose is mainly to demoralize, which was not the purpose in this case – and so the 'positive' side of the 'message' had to be determined.

Two decisions were taken on this, and both, as it turned out, had a fair measure of success. In the first place, it was decided that it would be wrong even to attempt to deceive the Japanese on the fact that there could be any exception for them or any other enemy power on the question of 'unconditional surrender'. At the same time Captain Zacharias, who had served in the U.S. Naval Attaché's office in Japan at various periods between the two World Wars, was to explain to his listeners that 'unconditional surrender' did not eliminate the application to Japan of the principles of the Atlantic Charter, one of the principles being the right of self-determination. In other words, the status of the Emperor and Japan's 'national structure' were, in the last resort, questions for the Japanese to sort out among themselves.

The second decision on the 'message' was that Captain Zacharias was to broadcast under his own name. After all, as a result of his various postings in the U.S. Naval Attaché's office in Tokyo, he knew many of the Japanese admirals personally, including Admiral Baron Suzuki who became Prime Minister of Japan early in April 1945. It was a deliberate policy decision to give his broadcasts an official imprint, and in one of his first broadcasts he explained that he was now going to read in Japanese a message specially drafted for him to transmit from the President of the United States to the people of Japan. The last sentence of the President's message read: 'Unconditional surrender does not mean the extermination or enslavement of the Japanese people.'

Captain Zacharias returned to the theme of the meaning of 'unconditional surrender' and its effect on the status of the Emperor and Japan's 'national structure' in broadcast after broadcast because he sensed that this was the nub of the argument between the 'peace faction' and the 'hard-liners' within Japan's power élite. Not all of the members of that élite may have heard every one of his short-wave broadcasts – reception, after all, was not always of the best quality, and it was easy to miss a word in an argument where every word counted – but Captain Zacharias and the OWI knew that a monitoring report, giving the complete text of every one of his broadcasts, would be circulated among the members of the power élite, both the 'hard-liners' and the 'peace faction'. Admiral Suzuki referred to Captain Zacharias' broadcasts obliquely in a broadcast he himself made, and one Dr. Kiyoshi Inouye, who was described as a commentator on international affairs, spoke of Zacharias in one of his broadcasts which he knew would be picked up by American radio monitors, as 'Zacharias-Kun'. In previous broadcasts Captain Zacharias had been referred to as 'Zacharias-Taisa' (Taisa for Captain). 'Kun' is a word usually used only among friends. In the highly stylized way in which the Japanese language is employed, the use of the word 'Kun' at this moment of time was not therefore without significance. Certainly, Captain Zacharias was hitting his 'target'.

And he continued to hit it after Nazi Germany's collapse in May 1945. He pointed out that the leaders of Japan's main European ally had preferred their country to disintegrate in chaos instead of even attempting to preserve its 'national structure'. Was that what Japan's leaders desired for their own country?

Slowly but surely, the 'peace faction' was gaining the upper hand within the Japanese power élite, but unfortunately not quickly enough. As the spring of 1945 dragged into summer without a definitive offer of surrender from the Japanese, and as American casualties mounted, President Truman felt justified in ordering two atomic bombs to be dropped on Japan. No doubt some historians will continue to argue that Japan would have surrendered, given a little more time; but that argument belongs to the realm of speculation. What is certain is that when Japan surrendered the 'peace faction' within the power élite ensured that there was no opposition to the landing of American forces and subsequently no attempt anywhere on the Japanese islands at any form of guerrilla warfare or sabotage. The same would not have happened if the counsels of men like General Tojo had prevailed. Captain Zacharias' broadcasts made a substantial contribution to the effective execution of Japan's surrender – a point which was amply substantiated by members of the power élite in the post-war years.

Another operation, initiated by OP–16W in close and harmonious co-operation with the OWI, was the broadcasts of Commander Norden to the German Navy. In this operation the objectives differed substantially from those that lay behind Captain Zacharias' project. By 1943 when Commander Norden began his broadcasts, the main activities of the German Navy were concentrated on U-boat warfare. Unlike the Japanese, the Germans had no fleets of warships to sally forth into the open waters of the ocean, to seek out enemy fleets and do battle for control of the seas. German submarines may have hunted in packs but it needed no admirals to co-ordinate and command these packs. As a result of these circumstances, Germany's admirals – unlike Japan's – did not play an important part in the inner councils of Hitler's Third Reich. They certainly belonged to no power élite with a significant influence on affairs. Admittedly, Hitler appointed Doenitz, an admiral, as his successor, but when he did that, Doenitz could do little more than initial a few documents of surrender, and in the case of the majority of surrender documents his initials were not even considered essential to validate the documents.

Germany's admirals, then – unlike Japan's – were not men to be courted but to be ridiculed. Ridicule was as effective a means of undermining the morale of German sailors, and especially submariners, as any. Hitler unwittingly contributed to its effectiveness because he was reluctant to see naval officers who were no longer fit enough on account of age or health to command U-boats, torpedo-boats or E-boats on active service, retire from the German Navy. Consequently, he promoted them to flag rank and spent much ingenuity on inventing commands for them. The reason for his propensity for promoting naval officers to flag rank was of course to some extent practical and perfectly rational: he needed senior officers in the Wehrmacht on whose personal loyalty he could rely; his army generals, he knew, tended to become somewhat disillusioned once they realized what heavy casualties his strategy was inflicting on the troops; the Luftwaffe generals were, on the whole, more imbued with Nazi ideology and blind loyalty to the Führer; and the same was true of the admirals, who, although they had no ships, owed their promotion, their privileges and not inconsiderable emoluments entirely to him.

During a visit to London, in the course of which he had detailed conversations with British Naval Intelligence, Lieutenant Commander Ralph G. Albrecht, U.S.N.R., became convinced that the German 'Admirals inflation' was a suitable subject to exploit. On his return to Washington he convinced OP–16W and the OWI of the plan, and he began broadcasting to German sailors in January 1943. His German was near-perfect, he was well-known as an international lawyer, and to hide his civilian background he was given the name of 'Commander Robert Lee Norden, U.S.N. or, in German, 'Fregattenkapitän Robert Lee Norden der Amerikanischen Kriegsmarine'.

In one broadcast he announced that he had just heard that 16 new admirals had been appointed in the German Navy. His figures were always accurate; as often as not they had been gleaned from German newspapers. That, Commander Norden continued, made a total in the German Navy of two grand admirals, five general admirals, and more than 150 other admirals. What, Commander Norden asked next, are all these admirals doing? Are they aboard their flagships? Unfortunately there were not enough warships in the German Navy for that. Since there were only 30 German warships of over 1,000 tons, five admirals would have to divide one flagship between them.

In April 1943 Commander Norden congratulated 20 German captains on their promotion to flag rank and added that Germany had that month produced more admirals than submarines.

In July 1943 Commander Norden informed his listeners that Vice-Admiral Lange had been appointed to take a command in the German Mediterranean fleet, the thirteenth flag officer in that theatre of war. It was a pity, he added, that the flagship of that fleet had been sunk on the very day he took up his post – a situation which posed quite a problem for him and his 12 fellow admirals.

In October 1943 Commander Norden reported that Admiral Schultz, the German naval commander-in-chief in the Crimea, had evacuated Sebastopol in a speedboat which flew his admiral's flag from the mast of the radio antenna.

In May 1944 Commander Norden estimated the number of warships in the German Navy to be between 90 and 100, and its admirals at 235.

It was small wonder that the Germans ceased to publish promotions to flag rank in September 1943. As several German sailors who were taken prisoner at about that time admitted, admirals had become a bit of a music-hall joke in Germany – like mothers-in-law. But Commander Norden tried to show more sympathy. After all, he said with just a tinge of sadness and fellow-feeling in his voice, was it not half the fun of being made an admiral to have all your friends, family and colleagues know about it?

Looking back over the wide and varied fields in which American propaganda, or psychological warfare, was active, the biggest obstacle it had to contend with was the lack of close, continuous contact with the top political leadership. It also suffered from organizational untidiness which inevitably caused friction, but this untidiness on the whole produced as many advantages as disadvantages. Although it drove some of the members of the various and frequently competing agencies to the verge of nervous breakdowns, it also made possible on occasion a degree of flexibility which might not have proved easy under a more rigid, family-tree type of bureaucratic organization.

The greatest benefit which accrued to the American propaganda machine was the consequence of historical accident and not of carefully planned design. It arose from the fact that America had to fight a propaganda war on two fronts. Germany and Italy on the one hand and on the other, Japan were two very different propositions with which to cope, and because the same agencies – mainly the OWI and the OSS – had to cope with both, they had to define their 'targets', the 'messages' they wanted to get across, and the 'means of communication' with the greatest care. Otherwise, their 'credibility' on one front or on both would have collapsed rapidly. Their work was the better for that challenge.

CHAPTER 12

Japan's Propaganda
Fight for its Cause

In July 1953 Commodore Matthew C. Perry, U.S.N., sailed into
Tokyo Bay with four American warships and ended two and a half
centuries of Japanese isolation from the outside world, but although
Japan rapidly transformed herself into a modern industrial state, she
remained very largely a closed society. It is a characteristic of closed
societies not only that it is difficult for outsiders to enter them and gain
even a measure of acceptance, but also that it is equally difficult for
insiders to obtain a clear picture of the world outside. The traditions,
the codes of behaviour, of conduct, of thinking, their closed society has
taught them, blur and distort their vision of the outside. This
imposed a well-nigh insurmountable handicap on Japanese propaganda
because it made it impossible for those in charge of that propaganda to
get their target into focus and so to devise a message that was meaning-
ful to that target.

The inability to think oneself into one's target's skin was nowhere
more apparent than in Japanese short-wave broadcasts to the United
States. There were four main themes which the Japanese pursued in
this field.

In the first place, they pointed out that America had no reason to be
at war at all. One Japanese broadcast contained this sentence: 'Not a
few of your compatriots feel bewildered at why America is at war,
America who could in tranquillity have enjoyed peace and prosperity,
menaced as she was by none at all.'

To anyone who was in the United States in the weeks and months
after Pearl Harbor, it must seem incomprehensible that the aggressor,
Japan, should seriously believe that her victim, the U.S.A., could be
convinced from Tokyo of all places that America did not know why
she was at war. Right up to the moment of Japan's surrender, many
Japanese appeared to have been completely blind to the fact that the
Japanese attack on Pearl Harbor was the only reason needed to justify
America's participation in the Second World War; no other was

required. These Japanese had to be taught a lesson; no other justification was necessary in the view of many Americans.

Plainly the Japanese misjudged the impact of Pearl Harbor on American public opinion. If they had been more flexible, they would soon have seen the error of their propaganda line by analyzing American newspapers and monitoring reports of domestic American broadcasts. No notice was taken of these analyses. The Japanese persisted in believing that Pearl Harbor would soon be forgotten, and that the majority of Americans would then be asking themselves what they were fighting for. It was Japan, Tokyo proclaimed proudly, that was fighting for the principles enshrined in the American Constitution, for the liberty and freedom of the enslaved peoples of East Asia whose cause was the same as that for which the Civil War had been fought.

No doubt the Japanese were quite sincere in pursuing this line. It was a far more positive and genuine form of propaganda than the rather sterile line pursued by the Nazis in 1940 and 1941 after their sweeping victories in Europe, when they talked about the establishment of a 'New Order'. Where talk of a 'New Order' always sounded insincere, hollow and unconvincing, liberty for the enslaved millions of East Asia had the ring of a genuine rallying call for a worthwhile cause.

The only trouble in directing it at the American target was that in the view of most Americans it came from a source which was tainted. And to have the Japanese proclaim themselves as champions of the principles of the American Constitution, was little short of sacrilege. A fundamentally false assessment of the target proved fatal to Japanese efforts in this line.

A second line the Japanese pursued in their propaganda against the United States was the decadence of the Americans. Materialism and good living had made the Americans soft. 'Why are you continuing to send your loved ones overseas to foreign countries to die for an unknown cause?', one broadcast began. The American people were not up to standing the strain of dead, wounded and maimed which war involved. They were not aware of the superiority of the Japanese fighting man, of his immense spiritual strength.

One wonders that no one in the Japanese propaganda services pointed out that this line of argument, addressed to an enemy, is likely to stiffen rather than to weaken morale. To compound their mistakes, the Japanese painted gruesome pictures of the hardships American families had to endure at home with ever-increasing food and clothes shortages. Japanese food rations were at that time barely above the subsistence level, and Americans who listened in to the Japanese broadcasts wondered whether the broadcasters were talking about conditions in their own country or in what they believed to be America. This type of broadcast undermined any measure of credibility Japanese propaganda may have tried to achieve.

The third line the Japanese pursued was the exploitation of what they saw as the stresses and strains in American society. They attacked Roosevelt as a 'Communist'; they asserted that the tax system favoured the rich at the expense of the poor; they became champions of the negroes and other racial and religious minority groups; they even espoused the cause of the rich. War profiteering and the exploitation of the workers by the bankers and the industrialists were constant themes.

In this sphere of propaganda the Japanese had done their homework, and done it well. When they attacked Roosevelt, they quoted verbatim from the Congressional Record on the debates on the New Deal. When they referred to racial discrimination, they cited the speeches of negro leaders. When they talked about war-profiteering, unfair tax-laws and the exploitation of workers by the employers, they read word for word excerpts from the speeches of trade union leaders. But while their diligent home-work did them credit, did they overlook the fact that differences in the American body politic and economic tend to be discussed and debated in somewhat robust language and that such differences sometimes go beyond discussion and debate to physical violence? The whole Japanese approach to this line of propaganda rested on a basic misconception of the nature of American society and the way it goes about its business, and it seems incredible in retrospect that anyone in Japan in the Second World War should have thought that the dissatisfaction felt by America's many minority groups could be turned into making them passive, let alone active, supporters of Japan's cause.

Perhaps the propaganda line which was potentially likely to yield the best results, was the fourth, the exploitation of the distrust many Americans felt towards their Allies, particularly the British and the Russians. Throughout 1943 and 1944 Tokyo reminded Americans that by fighting in Europe they were merely conquering another vast colonial empire for the Soviet Union. But this was an argument which the Japanese were careful not to push too far and which, although it was dropped into broadcasts now and then, was never pursued very consistently. The reason for this comparative reticence on what might have proved fertile ground, was not difficult to find. The Soviet Union was at that time neutral in her relations with Japan, and Japan, fully stretched in fending off the Allied counter-offensives on her empire, had no wish to provoke the Soviet Union into abandoning her neutrality and intervening in Manchuria. Consequently, Japan's reticence on this point increased in step with Russia's military successes against the Germans in 1943 and 1944.

Considerations of diplomacy did not compel similar reticence in Japan's propaganda towards America's other main Ally – Britain. 'Britain is proving,' one broadcast said, 'that Americans are the world's prize suckers. American troops are in England, India, Australia, Africa and New Guinea. They're there to save the British Empire.' Japan, in fact, was anti-Imperialist and anti-colonialist. So were many Americans who saw no reason why American soldiers, sailors and airmen should fight to restore Malaya to British rule, the Dutch East Indies to Dutch rule and French Indo-China to France. They did not approve of the detention in India of Gandhi, Nehru and other Congress leaders. President Roosevelt himself made no secret of the strength of his anti-colonialist feelings. In public he stated emphatically that the Atlantic Charter applied to 'all humanity'. In private he pressed Churchill for a fresh approach to India and Indian independence. In private, Churchill was unyielding though he did eventually send out the Cripps Mission to India which produced no results. In public Churchill declared at the Lord Mayor's Banquet in 1942: 'I have not become the King's First Minister in order to preside over the liquidation of the British Empire.'

It was a phrase the Japanese used over and over again throughout the rest of the War in their propaganda to the United States, to India and to the vast territories they had conquered in the Pacific and South East Asia. Roosevelt for his part eventually decided to stop pressing his anti-colonialist views on Churchill and made it clear to other Americans that he was not prepared to re-open the issue of Indian independence with Churchill. The two men agreed to disagree on this point. But when Lord Louis Mountbatten became Supreme Allied Commander, South East Asian Command (S.E.A.C.), and wanted to set up a joint Anglo-American psychological warfare division on the lines General Eisenhower had done in Europe, the Americans refused to participate because of their difference of approach to colonialism. American propaganda in South East and East Asia – apart from news and news comment – was concentrated on the Philippines, which had been semi-independent under the Americans before the War. The OWI broadcasts firmly committed the United States to giving the Philippines complete independence at the end of a war Japan was bound to lose.

Here then was a difference of approach between two Allies which, on the face of it, the Japanese should have been able to exploit. But they had no more success in the pursuit of this line than of any of the others.

The most likely reason for this lack of success was probably that most Americans simply did not believe what the Japanese told them. A contributory factor was that most Americans who took the trouble to listen in to English language broadcasts from Tokyo, found it difficult to follow what was being said. Short-wave broadcasts are not always easy to receive even under the most favourable conditions and with the most elaborate sets. There is always a risk of sudden unexpected interference or fading, however short. If on top of these difficulties the broadcaster speaks too fast, swallows syllables, uses un-idiomatic language or has a voice which is comparatively high-pitched, the result is bound to be an incomprehensible broadcast. As it so happened, many of the Japanese English-speaking broadcasters – even when they had been to American colleges – had all these faults. Unlike the Japanese-speaking OWI broadcasters, they were not carefully trained to speak slowly, clearly, distinctly and to use simple idiomatic language.

One of the exceptions to the comparatively low level of English language broadcasters from Japan was a young lady who became known as 'Tokyo Rose'. Her silky alluring voice turned many Americans in the Pacific and many British soldiers in Burma into devoted fans. 'Are you lonely,' she would begin, 'maybe your girl is out with some other boy.' Hardly a remark likely to raise the morale of a lonely, tired Marine crouching in rain-filled fox-hole. After the Battle of Leyte in the Phillipines she said: 'Orphans of the Pacific, you are really orphans now. With all your ships sunk, how do you think you will get home?' The men of Britain's Fourteenth Army in Burma, often called the Forgotten Army, were greeted with a caressing: 'Good evening boys. They say you are the Forgotten Men but I have not forgotten you. . . .'

Both American and British troops were apt to overlook her nastier asides. What they liked about her was her voice, the note of genuine concern in it and her taste in swing music. Some American bomber pilots admitted after the War that during one of their bombing raids on Japan they were planning to drop a batch of new records for 'Tokyo Rose' because the ones she was playing, were getting so scratchy. In fact, she became the nearest thing to a Pacific and Far Eastern 'Forces' Sweetheart – which can hardly have been what her masters in Japan's propaganda machine had in mind. On the contrary, the effect she had on her listeners appeared to have been the opposite of what was intended. As far as can be ascertained the only positive effect she seemed to have had which may have benefitted the Japanese was her campaign during intensive jungle fighting against the anti-malarial drug atabrine. She claimed that the drug made men impotent, and some U.S. Army doctors reported that as a result of these broadcasts a few men had not taken the drug as they should have done.

After the War 'Tokyo Rose' was discovered to be an American-born Californian of Japanese parentage, Mrs. Iva Ikuko Toguri D'Aquino, married to a Portuguese linotype operator on an English-language newspaper. She was arrested by the U.S. occupation forces as a war criminal but released after a year. Two years later she was arrested again, taken to the U.S., tried for treason and condemned to ten years' imprisonment. She was released after serving just over six years of her sentence. As a footnote to her story, it is worth recalling that the jury needed four days to reach a verdict and that some members afterwards openly expressed regret that they could not have decided differently.

If the propaganda effort of the Japanese against the Americans can hardly be called very effective, was Japanese propaganda in the vast areas she had conquered any more successful? Unlike the empty shell of a 'New Order', which was all the Nazis could present to conquered Europe, Japan – on paper at least – had something to offer: the Great East Asia Co-Prosperity Sphere, an end to white man's dominance, the liberation of fellow Asians in one huge union under the leadership of Japan. It was an exciting vision. The trouble was that it was a vision that could be realized only when the War was won, and to win the War, Japan needed the raw materials of the many lands in the Co-Prosperity Scheme – and needed them urgently. As a result the Japanese felt themselves compelled to exploit the lands they had conquered, far more ruthlessly and extensively than any of the colonial powers had done previously. Immediate short-term needs got in the way of and blotted out the long-term vision until there was nothing left to fire the imagination. To many people in the conquered lands it seemed that they had merely exchanged one form of foreign rule for another, and the new was crueller and harsher.

Apart from the policy of ruthless exploitation which the Japanese felt they had to pursue in the conquered lands to meet the demands of war, Japanese efforts to win the friendship of their fellow Asians were not helped by the behaviour of some of their troops. There were many cases of senseless brutality, but perhaps even worse from the point of view of trying to promote a sense of Asian brotherhood in the long run was the open contempt which many Japanese soldiers showed towards the populations of the conquered territories. Virtually nothing had been done by those in authority in Japan to persuade the ordinary Japanese that his mission was not only to conquer but that he was also the champion, the standard bearer of a cause that was to usher in a new era in Asia. But the concept of Japanese spiritual strength and superiority with which the Japanese had been indoctrinated was no preparation for this mission, and many Japanese looked upon the ways of the easy-going Malays and happy-go-lucky Indonesians with contempt.

Admittedly, the Japanese freed Sukarno whom the Dutch had imprisoned as a nationalist agitator, renamed the Dutch East Indies Indonesia and set up a native administration with which Sukarno co-operated. But Sukarno had few illusions about the Japanese, and when Japan started losing the War was quick to realize – like nationalist leaders elsewhere – that the real battle for the independence of his native land would be fought after the War and that the war period had to be used to prepare for that battle. In other words, there was a measure of co-operation between the Japanese and the nationalist movements. The two may have used the same language, but to both that language had a different meaning. Each used the other for its own ends but those ends were not identical.

A similarly ambivalent atmosphere characterized the relations between the various Indian 'independence' or 'liberation' movements in Japanese-occupied territory and the Japanese. By mid-1943 the most significant of these movements had become that led by Subhas Chandra Bose, a former President of the Congress Party and a disciple of Mahatma Gandhi. In 1941 this important political figure had mysteriously made his way to Berlin. In 1943 he disappeared from Berlin as mysteriously as he had arrived and some time later appeared in Tokyo. Within a short time he became the dominant Indian leader in Japanese-occupied South East Asia.

On 4 July 1943, at a mass meeting in Singapore, he announced the formation of a 'Provisional Government of India' and of an 'Indian National Army'. He became President and Commander-in-Chief, and in October 1943 made a formal declaration of war against Britain and the United States in the name of his 'Provisional Government'. For most of the rest of the War, however, 'Free India Radio' was the most aggressive instrument at the command of the 'Provisional Government', and Subhas Chandra Bose was its main speaker.

It was in his broadcasts and in those which the Japanese directed to India themselves, that subtle but significant differences of attitude and approach became apparent. Admittedly, the Japanese paid elaborate tribute to the bonds of Buddhism that united the two countries and continually stressed their staunch support for India's righteous cause, but the predominance in Tokyo of the 'hard-liners', the men who believed in the sword as providing the sole effective solution to Asia's dispute with the white man, was reflected in the tone and tenor of the

Japanese broadcasts. Tribute was paid to Mahatma Gandhi as the saint and leader of the Eastern Freedom movement, but in the respect that was lavished on the Mahatma there gradually crept a note of irritation, of cautious hints that in the world as it was, even so saintly a figure as the Mahatma might at some stage feel compelled to change his mind about non-violence. Eventually the director of the Nippon-India association, Fujima, went so far in a broadcast as to reproach Gandhi openly as being too idealistic. 'The present is no time for argument', Fujima said in a talk beamed to India. 'Only the sword will deliver India from her troubles, not mere words.'

Subhas Chandra Bose, with his vast knowledge of India and of the Mahatma's influence, went to immense pains immediately to dissociate himself and his movement from Fujima's comments. There may have been less than 200,000 radio sets in India capable of receiving short-wave transmissions, but Subhas Chandra Bose knew that those who owned and listened to those radio sets, determined Indian public opinion and could easily be offended by Fujima's views. Bose, therefore, continued to urge Indians in India to use methods of non-violence and to 'adhere strictly to our Mahatma's principles'.

Another factor militating against Japan's crusade against the white man and for a Great East Asia Co-Prosperity Sphere organized by Asians for the benefit of Asians was Japan's treatment of French Indo-China. The French colonial administration in Indo-China was pro-Vichy and had been given carte-blanche by the Vichy authorities to cope with the situation in the Far East as it saw fit. In practice this meant that the French in Indo-China gave in to every Japanese demand, and by the beginning of 1942 French Indo-China was virtually a Japanese colony, with French administrators little more than agents of the Japanese military. It may not have been a very glorious rôle for French military and civilian personnel, but the effect of this policy hurt Japan more than France. The Indo-Chinese and the Asians in the countries bordering on Indo-China found it hard to reconcile Japan's rôle as the champion of Asians against the white man with Japan's use of white men to run Asian territories for Tokyo's benefit. The reason for Japan's behaviour was of course simply that she did not have enough soldiers and administrators of her own to do the job. It was purely an arrangement of convenience, dictated by the grim necessities of war, but like all the other arrangements of convenience to which Japan was driven by the War, it vitiated the grand concept, the long-term vision of the Great East Asia Co-Prosperity Sphere in the eyes of Asian peoples whom she conquered for the short space of a few years.

Even if her 'message' to the conquered nations of East Asia had not been thus flawed from the outset, it would have been still further handicapped by the means available to put it across. After Japan's tremendous conquests in 1941 and 1942, Tokyo had the choice of either beaming its 'message' directly from Japan to all the conquered lands or of setting up propaganda machines in each of the conquered areas with broadcasting facilities of limited range. There were strong arguments in favour of each of the two solutions. Centralization would mean closer supervision of those in the conquered territories willing to co-operate with the Japanese; it would enable the Japanese propaganda machine to put the emphasis on the long-term ideal of an Eastern Asia ruled by Asians for the benefit of Asians and attempt to play down short-term hardships. Decentralization would bring about a sharper awareness of local difficulties than would have been possible from far-away Tokyo and so would be likely to contribute to a speedier solution in collaboration with local Japanese military and authorities.

The de-centralizers eventually won the day. The argument that brought them victory was that if the people in the conquered territories had radio sets capable of receiving Tokyo, they could also receive broadcasts from Japan's enemies. And as the tide of war began to turn, the Japanese certainly had no wish for any of the people in the terri-tories they had conquered to get even an inkling of the reverses their armed forces may have suffered. All radio sets in Indonesia, in Malaya, in the Philippines, in Indo-China and elsewhere, were called in and adjusted so that they were capable only of receiving local stations. Furthermore, the Japanese set up loudspeakers – 'singing trees' as they called them euphemistically – in village and town squares and made the local population assemble to listen together. Communal listening, the Japanese felt, discouraged the kind of brief disparaging remark which could so easily destroy the effect of a whole broadcast.

One great disadvantage of decentralization was that the quality of the personnel available varied considerably from one area to the next. It was also more difficult to supervise what local collaborators were doing. And worst of all, the local broadcasts soon lost sight of the vision of the future, the Great East Asia Co-Prosperity Sphere, and degenerated into appeals, not to say, threats, to produce and deliver more of the raw materials the Japanese war machine needed so desper-ately. In short, hardly the stuff to fire men's imagination.

The only area in which Japanese propaganda was successful was on the homefront; and there propaganda was probably needed least. Yet although Japan's power élite could count on the discipline which the traditions of Japanese society imposed on its members, the people were joined together in innumerable associations at the village, town and national level. The most important of all was the 'Imperial Rule Assistance Association', and new ones were formed all the time. There was the 'Student Physical Building Movement', the 'Special Summer Movement for Health and Morale Building', the 'One Hundred Million People Battlefield Spirit Promotion Movement'.

It is small wonder that the psychological warfare and intelligence agencies of the Allies at no time during the War regarded Japanese public opinion as capable of exerting the slightest influence on Japanese policy.

The home front apart, however, Japan's ineffective efforts in the Second World War provide an object lesson of how not to go about propaganda.

CHAPTER 13 Goebbels on the
Defensive

The Nazi propaganda machine to its own people, to the people of
occupied Europe and to the enemy countries was geared to military
success. Without military success it could not sustain its message that
the *Endsieg* (Final Victory) was close at hand, that the 'New Order' in
Europe was here to stay.

In the first half of 1942 Germany's propaganda machine had never
sounded more confident and sure of itself, and its tone had seldom been
as aggressive. The prospects for Germany and her Axis partners
appeared intoxicating. Germany had acquired an ally in Japan who had
already proved his worth, and with such a partner world conquest no
longer seemed a dream. The German armies in Russia would complete
the job they had not been able to finish because of the exceptionally
early onset of winter and would wipe out the remnants of the Red
Army and strike South, perhaps to link up with the Japanese in India.
Rommel would be reinforced in Libya preparatory to taking over
Egypt and the Middle East. America? America no longer counted. It
was America and not Germany that was trying to fight a war on two
fronts and making the painful discovery that she could hardly hold her
own on only one front.

Berlin told Britain that it was wishful thinking to expect American
aid; America, that she was sacrificing the flower of her youth and her
treasure in a vain attempt to uphold Britain's crumbling empire; and
occupied Europe, that Japan's entry guaranteed the permanence of the
'New Order'.

Against all this aggressive talk, Allied propaganda could only cry out
defensively: 'Hold on.' It could not even fully exploit Hitler's broken
promise of *'Endsieg'* in 1941. After all, there was no denying Hitler's
many victories during the Russian campaign of 1941, and the fact that
they were not as conclusive as he had promised could hardly be
balanced against the fall of Singapore or the loss of Burma. In any case,
Allied propaganda could not and did not claim that the Wehrmacht

had lost its aggressive bite. Everyone, including the Russians, expected a second massive German offensive as soon as the weather permitted in 1942, and the most anyone hoped for was that the Russians would somehow survive this onslaught. It was hardly an inspiring theme.

The summer of 1942 saw the peak of aggressiveness in German propaganda. In the last few months of 1942 the aggressiveness gradually evaporated, as the fortunes of war changed. In the autumn Rommel had penetrated into Egypt only to be defeated at Alamein by Montgomery. Despite their supposed absorption in the Pacific and the allegedly impenetrable barrier of the U-boats in the Atlantic, the Americans – less than a year after Pearl Harbor – had made a successful landing with the British in French North Africa. In Russia, at the end of Hitler's second summer offensive, the German VI Army had been encircled in Stalingrad. And in the Far East, Japanese military expansion had not only been contained but had begun to be pushed back.

For the first time a defensive note crept into German propaganda: 'Hold on. We cannot be defeated now after all the victories we have won, all the lands we have conquered. We cannot lose if we hold out.' It was hardly a line to win the hearts of a hostile occupied Europe. Nor was it particularly fortunate as far as the Germans were concerned because it stirred memories of the First World War when there had been a saying in German: 'We have killed ourselves by a surfeit of victories.'

Defeat or the prospect of defeat caught the German propaganda machine completely unprepared. In fact, its first reaction in some cases appeared to have been made up on the spur of the moment and bordered on the childish. 'Why', Berlin asked mockingly, 'have the Americans landed in French North Africa? There are no Germans there for them to fight. Perhaps that's the reason why French North Africa was chosen. It offered the opportunity of making a successful landing with no risk of becoming involved in any fighting. Or perhaps it was due merely to a navigational error.'

This kind of propaganda line revealed an absence of liaison with the military which had up to then been close. For while Berlin mocked the Americans landing in places far away from German troops, the Wehrmacht was landing men at airports around Tunis at the rate of 1,000 a day until eventually there were 150,000 German troops there.

Almost equally childish was the line German propaganda adopted over Montgomery's victory at Alamein over Rommel. 'What did it matter', Berlin said in a tone and manner which suggested that there were more important things to talk about, 'if Rommel, after the many shrewd and clever campaigns he had fought, was now forced to retreat before the overwhelmingly superior forces of Montgomery in Africa?'

The reactions on the part of Germany's propaganda machine to the first breath of defeat showed a remarkable measure of disarray. It seemed as though no one was in charge and one department did not know what another was doing. In the meantime, different departments played it by ear, and played it badly.

The story, however, does not end at this point. The Nazi propaganda machine may have been caught off-balance in later 1942 and early 1943 but the man in charge, Goebbels, was a professional in propaganda. While those working under him and those in Ministries like the German Foreign Office over which he had no effective day-to-day control, despite his close personal relationship with Hitler, may have stumbled from one mistake to the next, he set up what he considered to be the priorities in the propaganda battle in the context of a situation in which the Axis partners had plainly exhausted their offensive powers. And he sought to implement those priorities for the remainder of the Second World War.

The priorities were to keep up the morale, the fighting spirit of Germany, and where possible, of her Axis partners; to rally Europe to Germany's side against Communism; to convince the Allies of the formidable strength the Wehrmacht still retained; and to attempt to turn the Western democracies – as well as occupied Europe – against Soviet Russia. He pursued these objectives until the day in spring 1945 when he poisoned himself, his wife and his children; and until the day on which he committed suicide, he pursued them with energy and verve. Nazi Germany may have revelled in bar-room jokes about his club-foot and his peccadilloes with actresses; he was also certainly a very unsympathetic person; but he possessed a highly-trained and first-class intellect, he was a very hard worker who did not spare himself and worked astonishingly long hours, and to anyone who crossed swords with him in the propaganda battle he was a formidable opponent.

This was the man who, in the changed circumstances of the winter of 1942–3, pulled together the German propaganda machine when it was showing signs of fraying at the edges.

While his inferiors stumbled about helplessly in trying to explain away Rommel's defeat at Alamein and the Anglo-American landings in French North Africa and made laughing stocks of themselves, Goebbels concentrated on how to deal with a far greater disaster which he knew was going to be inflicted on the Wehrmacht shortly – Stalingrad.

In the case of Stalingrad there was no attempt by Goebbels and his propaganda machine to gloss over the magnitude of the disaster. On the contrary, Stalingrad was called 'The city of destiny', and the soldiers of the VI Army were compared to Leonidas and the Spartans of Thermopylae as well as to the Nibelungen, in the burning banqueting hall of King Etzel. Their struggle was 'heroic' and 'titanic', their mission 'historic', 'without parallel in the annals of mankind'. Goebbels recalled the glorious history of the VI Army, its triumphs in the Polish and French campaigns, its victories in Russia in 1941 and 1942. 'The VI Army', he announced as the end approached in Stalingrad, 'is not dead. It will never be dead. Long live the VI Army.'

The annihilation of an army of over 300,000 men was represented as a gigantic act of sacrifice to save Europe and Western civilization. 'Either Germany, the German armed forces and, along with us, our allies in Europe will win', Hitler declared in a proclamation after the surrender of Paulus, 'or the Central Asiatic Bolshevik tide will break in from the East over the oldest civilized Continent, just as destructively and annihilatingly as it did in Russia itself. Men's efforts, stretching back over several thousand years, to create a civilization would then have been in vain.'

Stalingrad – under Goebbels' direction – marked a complete and dramatic reversal in Nazi propaganda. Until Stalingrad, Hitler's Germany had been presented as a young and vigorous nation, impatient of the old ways and the traditionalism that held Europe back, eager to impose its 'New Order' everywhere. After Stalingrad, it became the self-appointed champion of Europe's ancient civilization against the threat of 'The barbaric hordes from the Steppes', the protector of the heritage of Athens and Rome, of Christianity, of European literature, sculpture, painting, architecture.

Until the end of the War this remained one of the dominant themes in Nazi propaganda, a theme directed at the Germans themselves, at the peoples of occupied Europe, and even at Russia's Western allies, America and Britain. 'Perhaps there are some clear-thinking people even in London', Goebbels declared, 'who have an idea of what it would mean if Europe lay at the feet of Bolshevism.'

Certainly in Germany this theme struck a responsive chord. Even as late as the spring of 1945, many German officers who could by no stretch of the imagination be described as pro-Nazi, were painfully surprised when, on being taken prisoners, their offers to fight on against the Soviet forces at the side of the British and the Americans were firmly rebuffed. They simply could not understand the refusal. Surely Communism was the common enemy.

The Wagnerian emotionalism which Goebbels whipped up over the loss of the VI Army was also made to serve a number of other purposes. It was used to divert attention from the fact that Hitler had been directly responsible for the Stalingrad *débâcle,* and it provided a smoke screen which made it difficult for ordinary Germans to get an idea of their country's true military predicament. Indeed, in an emotionally charged atmosphere in which the newspapers, the radio and the country's political and military leaders were constantly talking about what 'the noble heroes of Stalingrad' had done for the nation, it was considered positively sacrilegious, even among Germans who had no Nazi sympathies, merely to enquire whether it was still militarily feasible to continue the war.

'What', thundered Goebbels, 'are you doing to match the sacrifice of the VI Army?' This, at least as far as Germans were concerned, was a much more effective propaganda line than emphasizing that Germany had won too many victories and conquered too many lands to be beaten now.

Moreover, this line, in the months that followed Stalingrad, made it possible to impose new stringent controls of man and woman power, of rationing of foodstuffs and of the allocation of raw materials. Goebbels' language after Stalingrad may have been charged with emotion, but the emotion of his words never touched his own heart. His brain remained ice-cold. He knew exactly the effect he wished to achieve and the purposes he was seeking.

Germany's women, who – unlike those in Britain – had not up to then been subject to compulsory work, in offices or factories, were conscripted for work wherever the State said they were needed. Everyone, no matter how old or young, came under the order of Hitler's Minister of Labour. Food rations were reduced. All of Germany's resources were put on a 'total war' footing, and in occupied Europe the 'danger from the steppes', the 'Red menace', was made the excuse for raising still further levies of forced labour.

Above all, the Stalingrad *débâcle* and the threat from the East were used to play down the importance of setbacks in other theatres of war. Goebbels tried to make up for the errors made by his minions in dealing with the Anglo-American landings in French North Africa and Montgomery's victory at Alamein. He divided the various war fronts into those that were important and those that were secondary. The Eastern Front was important, and so could an Allied breach of the so-called Atlantic Wall be. Everything else, according to Goebbels, was merely secondary.

This division in the importance of the different theatres of war was a

Here...

Every camp has its doctors and its own well equipped clinic.

A football match in a German POW camp.

Left: Allied POW's making toys for their children to be sent home for Christmas via the International Red Cross.

Right: Last Christmas every room in the German camps had its mistletoe or a Christmas tree.

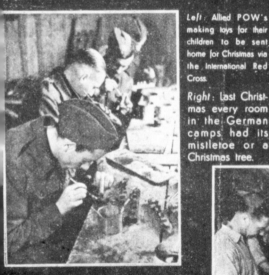

Right: Once a week there are the welcome Red Cross parcels from home. On that day you see nothing but smiling faces.

LIFE IN A GERMAN POW CAMP

15. An attempt to persuade Allied soldiers to surrender and survive.

16. The Roosevelt quotation was irrelevant when dropped on Anzio in 1944, and therefore without impact.

The Girl You Left Behind

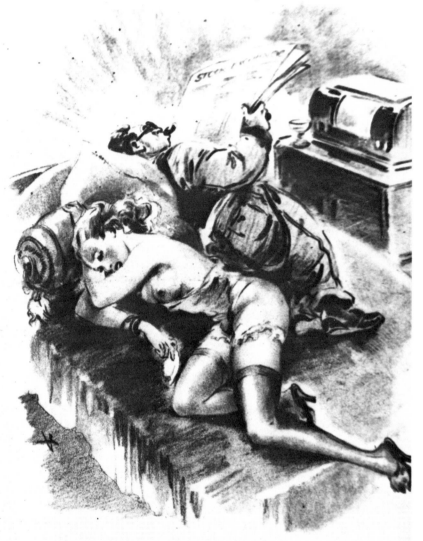

Poor little Joan! She is still thinking of Bob.......

17. A theme which never failed to stir up unease among fighting men.

18. A poster aimed at Russia's millions: its stark inscription reads 'Avenge'.

not unimaginative ploy in the propaganda war. Most Germans – whether or not they were Nazis – agreed with Goebbels' argument; in the minds of the people of Europe, it stoked the embers of suspicion that the Anglo-Americans were not doing as much as they might; and it helped to increase Soviet resentment at what the Russians regarded as undue and unjustifiable delay on the part of the Anglo-Americans in launching the 'Second Front'.

Of course German arms had suffered reverses in Africa, Goebbels argued. But Africa was on the periphery of the War. The War would not be decided there. And the surrender in Tunis of an Axis army of over a quarter of a million men, half of them Germans to the Anglo-American forces under General Eisenhower? Regrettable, yes; but, seen in the long term, merely a sideshow, an effort to buy time for the battles that mattered. Throughout the North African campaign news of the fighting there was treated as irrelevant to the essential prosecution of the war and tucked away at the end of radio bulletins and on the inside pages of the newspapers – if it was mentioned at all.

Even Italy, when the Allies invaded it, was dismissed as a *'Vorfeld'* – as outlying terrain not vital to the security of what the Germans now came to call *Festung Europa* or Fortress Europe. Mussolini's fall from power, the armistice which Italy concluded with the Allies and, later, Italy's co-belligerency, these gave Goebbels little trouble as far as German morale was concerned. Most Germans had never taken the Italians very seriously as allies anyway, and when the German army, without meeting any serious resistance, moved in with lightning speed to take control of the parts of Italy not occupied by the Allies and at the same time stabilized the front against the Allies, Goebbels congratulated the German armed forces on their success but continued to insist that Italy was merely a sideshow.

Goebbels' constant theme was that it was on the Eastern Front that the War would be won or lost, and that the only place where the Allies could strike a telling blow and crack *Festung Europa* was through the Atlantic Wall. The Russians, Goebbels went on to explain, knew this, and that was why they were furious with their Western Allies for allowing hundreds of thousands of their soldiers to be pinned down in Italy by a handful of crack German units who made them fight every inch of the way in battles that were of no consequence. And why did the Western Allies not fight where it mattered? Because the Atlantic Wall was impregnable. Its fortifications were being continually strengthened. It was manned by some of the best units in the German Wehrmacht and led by Germany's ablest commanders.

Throughout the winter of 1943–4 German newspapers were full of photographs of the craggy features of Field Marshal von Rundstedt, perhaps Germany's best general and certainly the one who inspired the greatest confidence in the German public, against a background of forbidding-looking pillboxes and tank traps on the French coast. And the appointment of Rommel, Germany's most popular general, to a high command under Rundstedt gave the Nazi propaganda machine the opportunity of still further bolstering the German public's wish to feel confidence in the impregnability of *Festung Europa*.

Thus, having been caught unprepared by the first wave of reverses to German arms, Goebbels rapidly devised a coherent defensive strategy for German propaganda. And within the broad framework of that strategy he still scored the occasional 'propaganda triumph'. Perhaps his greatest was his handling of the Katyn massacre.

The horrifying story of Katyn began with the Nazi-Soviet pact of 1939. According to the terms of that pact, after Germany had invaded Poland and crushed Polish resistance approximately one half of Poland was occupied by Russian troops and such parts of the Polish army as happened to be stationed in and managed to escape to Russian-occupied Poland, were interned in three Russian camps. After the Nazi attack on Russia in 1941, the Soviet Government – as a result of strenuous efforts made by the British government – established diplomatic relations with the Polish government in exile in London and signed an agreement with that government to release all Poles interned on Polish soil. Most of the prisoners in one of the three Russian camps were either released or recruited to form a new Polish army, but the prisoners in the other two camps – about 15,000 officers and men in all – were never located. All enquiries about their fate by the Polish government in exile were met with the reply that the Soviet authorities knew nothing about these officers and were unable to account for them. General Sikorski, the Polish Prime Minister in exile, eventually decided to travel to Moscow in person, and he saw Stalin and Molotov to attempt to solve the mystery of the missing prisoners. But the two Soviet leaders, like their officials before them, professed total ignorance of the 15,000 prisoners. On General Sikorski's return to London, then, the position was that the Russians claimed that all Polish prisoners had been released while the Polish government in exile maintained that 15,000 officers and men who had been in contact with the outside world until the early spring of 1940 (but not thereafter) when the Russians still controlled the area around Katyn, were missing and unaccounted for.

General Sikorski next turned to the British Government for help and persuaded it to intercede with Moscow on Poland's behalf. Britain brought the strongest possible pressure to bear in Moscow, but in the end she was no more successful than the Polish government in exile. Despite Britain's failure, the Poles persisted in their enquiries, but each enquiry continued to be met with a bland profession of ignorance. And needless to say, throughout all this time relations between Moscow and the Polish government in exile steadily deteriorated.

London and Washington watched this situation with growing concern. Both the British and American governments knew that much more was at stake than relations between the Polish government in exile and Moscow. A rupture between the two could have much more far-reaching results affecting relations between the Soviet Union and the Western Allies. Relations between the Polish government and Moscow were the weak link in the Soviet-Anglo-American alliance, the link which – if it broke – could in certain circumstances jeopardize the alliance itself.

Britain had, after all, gone to war in 1939 to preserve the autonomy of Poland. Understandably, therefore, she had worked hard and patiently to bring Moscow to recognize the Polish government in exile and she was making strenuous efforts to see that diplomatic relations between the Polish government in exile and the Soviet Union continued to be maintained. In the British government's view, it was only through the Polish government in exile, a government recognized and accepted by Moscow, that there was any hope of re-establishing an autonomous independent Poland after the war. This view was shared by Washington, which was as much a champion of the Polish government in exile in London as the British government. After all, the right of self-determination, enshrined in the Atlantic Charter, applied, in Roosevelt's words, to 'all humanity', and it was inconceivable that one of the first victims of Nazi aggression should be denied that right. Apart from these broad strategic and idealistic considerations, there were also internal political pressures inside the United States which helped to strengthen support for General Sikorski's government in London. America contained large Polish minority groups who were grimly anti-Russian and were not prepared to countenance any arrangement agreed to by the U.S. which would replace German rule over Poland by Russian rule. As it so happened, these minority groups lived in areas where their vote could make the difference between victory or defeat for one political party or the other.

Polish-Russian relations could therefore well become a major point of disagreement between the Allies. Indeed, in the view of many British and American officials who had to deal with this delicate situation from day to day, there was enough explosive material lying around to blow the alliance sky-high. Fortunately, the whole matter had been kept completely secret. The most elaborate precautions had been taken to ensure secrecy. The public knew nothing of the 15,000 missing Polish prisoners, of General Sikorski's interview with Stalin and Molotov, of the increasingly bitter exchanges between the Polish government in exile and Moscow, of Britain's efforts to intercede. So long as secrecy could be maintained, there was still hope. Once the story became public – as all those involved knew and feared – anything could happen.

That was the situation in April 1943 when Goebbels, who, through his intelligence services, knew the background, decided to light the fuse and see what the explosion would produce for him to exploit.

On 13 April 1943 Berlin Radio carried this news item: 'It is reported from Smolensk that the local population has indicated to the German authorities a place in which the Bolsheviks had perpetrated secretly mass executions and where the G.P.U. (The Soviet Secret Police) had murdered 10,000 Polish officers. The German authorities inspected the place . . . and made the most horrific discovery. A great pit was found . . . filled with . . . layers of Polish officers. . . . Many of them had their hands tied, all of them had wounds in the back of their heads caused by pistol shots. . . . The total figure of the murdered officers . . . would more or less correspond to the entire number of Polish officers taken as prisoners of war by the Bolsheviks.'

The secrecy which had surrounded the tragic story of the missing 15,000 Polish officers and men, had been destroyed and Goebbels could afford to sit back to see the damage his explosion had caused, and then, once the dust had settled, exploit the situation as he saw fit. He did not have long to wait.

The Polish government in exile in London confirmed the Berlin broadcast that 15,000 Polish officers and men, imprisoned by the Russians, were missing and that the Russians had refused to account for their disappearance. The Russians denied any connection with the atrocity and accused the Germans of murdering the Polish officers themselves after their conquest of the Smolensk area. This accusation had not even been hinted at in the secret exchanges between the Polish government in exile and Moscow, or in the exchanges between Moscow and London to Poland's behalf. It therefore served only to increase the suspicions of Polish exiles in London.

This was the moment Goebbels chose to try to exacerbate the situation a little further. His broadcasts gave not only lurid details of the scene of the burial site in the Katyn area but also the names and ranks of the bodies as they were being recovered. And Berlin promised to give more information on identification as the bodies in this tragedy were recovered hour by hour. 15,000 men have a lot of friends and relatives, and a lot of those relatives and friends were in Britain, in America and in North Africa. The pressure of Polish public opinion on General Sikorski's Polish government in exile for a showdown with the Russians, regardless of the consequences, was tremendous. All that American and British news and propaganda services could do, was to broadcast and print the Soviet Union's denials and the counter-charges against the Germans. They did so without making any comment.

The Polish leaders in London eventually decided that the only way to ascertain the truth was to examine the evidence at the burial ground at Katyn. They appealed to the International Red Cross to conduct such an investigation.

Goebbels immediately took up the Polish request to the International Red Cross and announced that Germany would give an International Red Cross investigating body every facility to conduct its enquiries.

After the Polish and German requests to the International Red Cross, *Pravda* attacked the Polish government in exile as a false and treacherous ally in an article which bore the head-line 'Hitler's Polish Collaborators'.

The International Red Cross, for its part, refused to undertake the investigation on the grounds that it was unable to get the consent of all the parties involved. This was perfectly in accordance with the procedures of the International Red Cross, but since the Poles and the Germans had requested the investigation, the world at large was left with the impression that it was the Russians who had opposed it because they had something to hide. It is an impression that persists to this day, and there are many – especially in Britain and America – who are firmly convinced that when it comes to atrocities and disregard for human values, there is nothing to choose between Nazi Germany and Soviet Russia. The Western Allies, they argue, should have taken a more positive line with Moscow instead of merely publicizing its denials and counter-charges against the Germans without making any comment.

In the meantime Goebbels, while publicly regretting Russia's refusal to participate in an International Red Cross investigation, announced Germany's intention of establishing its own independent commission to investigate the Katyn crime. It included specialists from all the West European Continental countries, including Poland and Switzerland, and it confirmed the German allegations. Reports of the commission's findings and photographs of the appalling scenes of the mass graves in woods around Katyn were carried in neutral newspapers throughout the world.

All the while, the inexorable logic of ancient hatreds, mistrust and present animosity was grinding on relentlessly. Less than a month after the first Nazi broadcast on Katyn by Berlin Radio on 13 April 1943 the Soviet Union broke off diplomatic relations with the Polish government in exile in London. Shortly afterwards, Moscow set up a 'Committee of Polish Patriots' which was later to form the nucleus of a Russian-sponsored government as a rival to the Western Allied-sponsored Polish government in London.

Goebbels' propaganda had been shrewd and effective. He had assessed the psychology of his various targets correctly, and by the content and form of his message set one against the others. He was probably not, however, as effective and destructive as he may have hoped to have been; for although he may have forced Stalin to reveal his aims about dominating Central and Eastern Europe earlier than the Soviet leader would have wished, did Goebbels really imagine that that was not Stalin's purpose in the long run? And, again, although he caused a rupture in Polish-Russian relations, he failed to disrupt the grand Soviet-Anglo-American alliance. In 1943 the Soviet Union and the Great Powers of the West needed one another too badly to be able to afford the luxury of falling out.

Yet although Goebbels may not have achieved all he had hoped for from Katyn, it represented one of his greatest triumphs in propaganda. It was also his last. In constructing the strategic framework for Nazi Germany's defensive propaganda, Goebbels had laid down that there were war fronts on which the War was going to be won or lost – the Eastern Front and the Atlantic Wall – and the war fronts on the periphery which were essentially irrelevant to the final outcome.

It was a not unreasonable scheme – at least until it became eroded by military reality. It became virtually meaningless by the time the Western Allies had breached the Atlantic Wall and broken out of the Normandy beachhead, and the Russians were rolling across the Baltic countries and pushing into Poland, Hungary and the Balkans. Even as resourceful a propagandist as Goebbels needed something on which to exercise his ingenuity, and by the winter of 1944–5 there was little, if anything left.

As a result Goebbels' propaganda, to all intents and purposes, took off into what can be described only as the outer space of psychological warfare. He talked of 'miracle weapons' – which Germany was developing and which would yet snatch victory for Germany from the jaws of defeat at the last minute. And when Roosevelt died in April 1945, he told the German people exultantly that he had always predicted that Germany would be saved at the eleventh hour by an unexpected event. The same thing had happened almost two centuries ago in the sixth year of the Seven Years War in 1762 when the Czarina Elisabeth, the ally of Austria and France, had died unexpectedly and been succeeded by her half-witted nephew, Peter III, an ardent admirer of Frederick the Great, who had immediately called off Russia's war with Prussia and so ensured Prussia's survival. Roosevelt's death would have exactly the same effect as Elisabeth of Russia's, Goebbels concluded triumphantly, because Roosevelt had been Germany's arch enemy and without him the alliance would collapse.

Reports have it that Goebbels was so carried away with his own rhetoric that he ordered champagne to celebrate and telephoned Hitler in his bunker to tell him that this was the turning point. What effect his broadcast had on the Germans who had time to listen to him, was doubtful. With the Western Allies well across the Rhine and the Russians in East Germany, they may have found the relevance to present day circumstances of his historical analogies of events that had taken place nearly 200 years earlier a little difficult to appreciate, and some may even have wondered if Harry S. Truman, the man they were told had taken over the American Presidency, could be compared to a half-witted Peter III, who even by the standards of eighteenth-century Czarist Russia was considered so unpredictably unstable that he had to be done away with.

In any case, propaganda at that stage of the War was irrelevant. It was no longer a question of sustaining morale and the fighting spirit. Whether German formations fought on or surrendered depended mainly on the presence or absence of S.S. or Gestapo units in the vicinity. German defences, like Germany's defensive propaganda, had effectively collapsed.

CHAPTER 14

The Allies on the
Offensive:
'White' and 'Black'
Propaganda

While a defensive note crept into German propaganda in the last few months of 1942, Allied propaganda went over to the offensive. Echoing Churchill, it drove home to friend and foe alike: 'This is the beginning of the end for the Axis.' And its main task, now that victory beckoned, was to weaken the enemy's will to resist and so to shorten the War.

The main purpose of Allied propaganda after Stalingrad was to dispel the cloud of Wagnerian emotionalism with which Goebbels loved to smother all critical thought inside Germany about the military situation of the country – in short, to bring the Germans down to earth and face facts. From the West this task was tackled by the B.B.C. and the Voice of America, and from the East by Radio Moscow and a series of so-called 'freedom stations' under Russian control.

If more Germans, by all accounts, listened to the B.B.C. than to any of the others, this was probably due in no small measure to the fact that London was Berlin's oldest enemy. Unlike Moscow and Washington, Britain and Germany had been in the War from the beginning. London was not comparatively new to the game like Washington, nor felt to be as prejudiced and biased in its views as Moscow was presumed to be, as a result not only of the dictates of Communist philosophy but also of the bitterness and brutality of the fighting on the Eastern Front.

Certainly London's memory went back over the years, and the B.B.C. used that memory to good effect. When Goebbels called for sacrifices from the German people after Stalingrad, the B.B.C. transmitted recordings of speeches by Hitler in which he celebrated his triumphs of 1940. While the German radio spoke of the noble and heroic self-sacrifice of the VI Army, the B.B.C. gave casualty figures followed by the recording of a speech in which Hitler assured the Germans, in the autumn of 1941, that the Soviet Union 'is already destroyed and can never rise again'. When Goebbels explained that

Hitler had conquered so much ground that he could afford to give up a lot without fear of defeat, the B.B.C. broadcast a recording of Hitler's speech promising *Endsieg* (final victory).

While Goebbels tried to hide the Führer's strategic mistakes from the Germans, the B.B.C. made Hitler condemn himself with his own voice. It also set about exploding Goebbels' strategic explanations. When he dismissed the capitulation of the Axis forces in Tunis as 'peripheral', the B.B.C. commented that 130,000 Germans plus 130,000 Italians were a lot of people, that they had a lot of relatives and friends. If this was what happened in a 'peripheral' theatre of war, what could the Germans expect in a central, vital theatre of war?

London, like Washington and Moscow, never for a moment allowed the Germans to forget the truth of their country's worsening military situation, and as the German news about the Eastern Front became progressively more vague and evasive, so the Allies became more pointedly precise. The reputation the B.B.C. had gained for accuracy and honesty of its news in the grim days of 1940 and 1941 stood it in good stead at this stage. When the Germans announced that their Army Group Centre had straightened its lines according to plan – without mentioning any place-names – London pointed out that the Germans had been pushed back over 100 miles in the first days of a new Russian offensive and that Smolensk was expected to fall shortly. It fell two days later. When the Germans talked of heavy, heroic fighting against overwhelming odds on the North Russian front, the B.B.C. told its listeners that the German siege of Leningrad had been broken. And when the Germans announced that the Ukrainian front had been stabilized, London commented that the Ukrainian front was now in Rumania.

German listeners – and their number was increasing constantly as hard news from German sources became sparser – were not even allowed to console themselves with cosy thoughts about comparative security within Hitler's *Festung Europa*. London was more closely in touch with the Resistance movements in Occupied Europe than any of the other Allied capitals, and left the Germans in no doubt that their fellow inhabitants of *Festung Europa* did not see them as champions of Europe against the barbarian tide from the East; they would indeed seize the first opportunity to revolt. At the same time London constantly counselled restraint when addressing the peoples of Occupied Europe: don't show your hand too soon; that is what the Germans, who are becoming increasingly nervous and jumpy, want you to do so that they can crush you and eliminate you; so wait until the decisive moment; we will tell you when and how your help can be most effective.

Allied efforts to undermine enemy morale were not confined to 'white' propaganda, to stations like the B.B.C., Radio Moscow, or the Voice of America which were clearly identifiable and wanted listeners to recognize who they were and what they stood for. There was also a considerable effort in the field of 'black' propaganda from stations which pretended to be what they were not.

In 1939 and 1940 the Germans had operated a number of stations from inside the Reich which tried to convince listeners that they were being run from inside France by 'patriotic' Frenchmen who were disgusted by the treachery and corruption of the leaders of the Third Republic. Later on they had for a time operated 'Radio Free Caledonia', allegedly run from secret hide-outs in the Highlands by sturdily independent Scotsmen, eager to rid their land of the English oppressors. But German attempts at 'black' propaganda were small and insignificant compared to what the Allies did in the second half of the War.

Of all the 'black' stations launched by the Allies, the most ambitious and effective by far was the 'Kurzwellensender Atlantik' ('Short-Wave Station Atlantic') which began operating in February 1943 and was later, in October 1943, expanded into the 'Soldatensender Calais' (Soldiers' Station Calais), broadcasting not only on short but on medium waves. Both pretended to be German forces broadcasting stations, operating from somewhere inside France and serving all the personnel of the various German army, Luftwaffe, and navy commands stretching from Norway down to the Bay of Biscay – though in fact both were run from inside Britain by Britain's Political Warfare Executive.

The man who conceived and directed these 'black' propaganda operations was Sefton Delmer, a tall, big man who during the War sported a beard. As a result the Americans of OSS with whom he worked from time to time – like General 'Wild Bill' Donovan – occasionally referred to him as 'The Beard' or 'Henry VIII'. During the black-out in London he was sometimes also mistaken for Ernest Hemingway. Before the War he had been one of the top foreign reporters on Lord Beaverbrook's *Daily Express*. He spoke French and German perfectly. Indeed, if he wanted to, he could adopt the exact intonation and phraseology of a Berlin taxi-driver, and native-born Germans were frequently left wondering if he was not having them on when he told them he was British. He had watched the events in Europe during the 1930's leading up to the War at first hand. He had seen the succession of crises in France which had weakened the foundations of the Third Republic. He had covered the Spanish Civil War both from the Republican side and from the Franco side. Above all, he had observed the collapse of the Weimar Republic and the rise of Hitler. In 1932 and 1933 he had accompanied Hitler on his campaign tours all over Germany. He knew the members of the Führer's entourage and principal lieutenants, their personalities, their strengths, their weaknesses, their idiosyncrasies. Never was there a band of men who disregarded even the most elementary rules of security with such abandon. They talked and talked among themselves about their plans, their ambitions, their methods and about each other, and Delmer was present at many of these exchanges. In short, he possessed one of the essential qualifications for a propagandist – a knowledge in depth of his target. He also had the ability to think himself into the other fellow's skin.

The first 'black' propaganda operation launched by Delmer, was a short wave station which called itself 'Gustav Siegfried Eins'. 'Gustav Siegfried' was the German signallers' equivalent of 'G for George' and 'S for Sugar'. It was therefore highly mysterious, more so as 'Eins' (One) was added for good measure, and to add to the puzzlement and confusion of the German radio monitors and security services, the broadcasts always included coded messages which, when deciphered, told the security services that mysterious messages had been sent by 'Gustav Siegfried Eins' to 'Gustav Siegfried 12' or 'Gustav Siegfried 18'. The suspicion was thus sown in the minds of the German security chiefs that 'Gustav Siegfried Eins' might be sending messages to a network of British agents in Occupied Europe. German security, with the technical detection at its disposal, quickly discovered that the broadcasts originated in England and not from a mobile transmitter in a van somewhere on the Continent which just managed to keep a step or two ahead of the Gestapo. Simpler souls, however, who believed the story of the elusive 'Gustav Siegfried Eins', may have felt that there really was an underground network stretching right across Occupied Europe and plainly including members in the higher echelons of the Wehrmacht.

The central character in the 'Gustav Siegfried Eins' broadcasts was a man who was introduced as 'Der Chef' – 'the Chief'. Delmer chose that description for him because in 1932 and 1933 he heard Hitler referred to by his immediate entourage, not as the 'Führer' but as 'Chief'. Delmer's 'Chef' was loyal to the Führer. The 'Chef' was patriotic, nationalist and a hard-line German chauvinist. By calling him 'Der Chef', it was not necessary to identify him in Nazi-Germany's hierarchy. His language, accent and delivery of what he had to say, did that much more effectively. He was plainly a die-hard, obstinate, patriotic reactionary Prussian officer of the First World War who knew his way around and held a fairly senior position in the Wehrmacht – a position which no one could define accurately; but since he was a man of authority and with inside knowledge, it was obviously worth listening to him. And Delmer made sure that German General staff officers as well as dedicated Nazi Party officials and members of the German public who were wondering – once the tide of war turned against Germany – what was really happening and wanted reassurance from 'professional soldiers' they trusted, recognized that 'Der Chef' was the genuine article.

Delmer coached his 'Chef' to adopt that slight trace of a Berlin drawl which he had found among the Prussian nobility, the 'Junkers', who provided the majority of the officers in the Kaiser's Guards regiments. He also provided him with an adjutant. This was important because Delmer had noticed that not even half-way senior army or Nazi Party officials ever appeared in public without at least one adjutant in attendance, more if possible. It did not seem to matter that this practice involved a tremendous waste of manpower. What mattered to Delmer was that a man as senior as 'Der Chef' obviously was, could not appear even on a clandestine broadcast without at least one adjutant. And so Delmer found one who announced 'Der Chef' with correct military curtness and conviction. He used four words only: *'Es spricht der Chef'* (Here is the Chief).

Delmer also provided 'Der Chef' with a signature tune. Hitler's Deutschlandsender's signature tune was the first few bars of an eighteenth-century folksong as played on the bells of Potsdam's famous garrison church where many of Prussia's kings, including Frederick the Great, had attended religious services. The words to those few bars were:

> *'Ub immer Treu und Redlichkeit. . . .'*
> ('Always practice truth and probity. . . .')

Delmer chose the second line of that folksong as the signature tune for 'Gustav Siegfried Eins'. The words of that second line were:

> *'Bis au dein kühles Grab. . . .'*)
> ('Until your cool, cool grave. . . .')

And he had the tune of that second line played on what seemed to be a cracked piano in some primitive front line trench.

'Gustav Siegfried Eins' was launched well before the tide of war turned in the Allies' favour. The first broadcast was on 23 May

1941 – three weeks after Hess, Hitler's deputy, parachuted down in Scotland, and less than a month before Hitler was to invade the Soviet Union.

The B.B.C. gave merely the bare news of the parachute descent of Hess. There was no comment, no speculation about its possible or potential significance.

'Gustav Siegfried Eins' – 'Der Chef' – in his first broadcast, which incidentally was not announced as the first but as one in a series which had been going on for some time in order to cause the greatest possible confusion in Germany's security services and the maximum amount of recrimination inside Nazi monitoring organizations, did not feel similarly inhibited. That first broadcast revealed how Delmer was to exploit the German 'nationalistic patriotic line' in later years when the Allies were gaining the upper hand militarily.

'Der Chef' placed Hess firmly among the clique of cranks, megalomaniacs, string-pullers and parlour Bolsheviks who formed a circle around Hitler and were in effect Germany's real leaders. Hess simply did not have the nerve to stand up to a crisis. Once things went wrong, he dived for his bag of hormone pills and a white flag and flew off to throw himself on the mercy of the unspeakable British. Unspeakable is indeed an understatement. 'Der Chef' used the foulest and most obscene language that has been transmitted over the airwaves before or since.

After seeking to destroy Hess's character and credibility, 'Der Chef' went on to reassure his listeners on one point. Rumours that Hess flew to Britain on the Führer's orders were quite untrue. The Führer, 'Der Chef' stated categorically, would never have authorized a man with such an intimate knowledge of Germany's military operational plans to go enemy country. Unfortunately that obscenity of a Reich Security Chief, one Himmler, had allowed Hess to get away. Of course the generals were right to query the feasibility of what was going to happen. Their only mistake was to tell the obscenity of a so-called Deputy Leader. But then with so many lick-spittles about, who could get anywhere near the Führer to tell him the truth. And in the meantime, Himmler was arresting good German 'patriots' who should be allowed to tell the Führer what was going on, in order to cover up his own mistake in allowing his useless snoopers to let Hess get away.

To rub in the point that 'Der Chef' was going to be around for some time, he ended his first broadcast with the sentence; 'I shall be repeating this – all being well – every hour at seven minutes to the full hour. *"Immer sieben Minuten vor voll"*.'

This first 'Black' broadcast under Delmer's direction reflected the objectives he had set himself. In the first place, to split the Nazi Party 'lick-spittles' from the Führer and the more conservative elements in the German army; secondly, to shake the confidence of the German High Command and public, in their own security – it must have been highly unpleasant to be told that Hess knew all the army's operational plans less than a month before the Nazi invasion of the Soviet Union, a ploy which with variations Delmer was to exploit again and again in the years that followed; thirdly, to portray the Nazi Party 'lick-spittles' as self-seeking obscenities, determined to feather their own nests, regardless of the higher interests of the Fatherland; and fourthly, to portray the ordinary German soldier, sailor and airman as honest, courageous and patriotic – unlike some Party members for whom Delmer and his colleagues invented the imaginative collective term *'Parteikommune'* – 'Party Commune' a term which somehow suggested that Nazi Party members – especially high-ups like Himmler – were partially infected with the creeping disease of Communism.

To confuse the German security forces still further, 'Der Chef' in his first broadcast spoke of Hess throwing himself 'on the mercy of that flat footed bastard of a drunken old Jew Churchill'. Himmler's security services puzzled over this phrase for many a week. Their technical radio experts may have assured them over and over again that the signal of 'Gustav Siegfried Eins' came from England, but could even a 'black' station use that kind of language about Britain's war leader?

In the months that followed, 'Der Chef' busily spread rumours to undermine morale, or 'sibs' – as they were called from the Latin *sibillare,* to whisper. Any general appointed to high command, especially on the Eastern front, was likely to be a 'lick-spittle', unfit to command a platoon. All the blood transfusions given to German casualties from the Eastern front came from Russian and Polish donors; since none of these donors had been subjected to the Wassermann test, was it surprising that venereal disease had broken out in a number of field hospitals on the Eastern front?

'Der Chef' built his 'sibs' on first-class intelligence. One such source of intelligence was the American-born wife of a Cologne industrialist. She wrote to her friends in America about the social life she was enjoying in Cologne amid such Nazi luminaries as Gauleiter Grohé and Cologne's Chief Mayor, Herr Winkelkämpner. Her letters were intercepted, and used by 'Der Chef' in his campaign against the *Parteikommune.* Of course, 'Der Chef' – under Delmer's directions – embellished the contents of the letters a little, but there was enough truth in what he said to make public contradiction impossible. In any case, the German public was fast reaching the point where it was willing to believe the worst of Nazi Party officials. That obscenity of a Cologne Mayor, 'Der Chef' noted caustically, had recently given a party at which his guests had been treated to a cake shaped like Cologne Cathedral. Cologne cathedral was a vast edifice by any standards, and so the cake had required kilos of sugar to make; funny – 'Der Chef' continued – in view of the fact that the sugar ration had just been cut drastically for ordinary mortals. But then why should a member of the *Parteikommune* know a small detail like that?

Needless to say, 'Der Chef' did nothing to improve relations between Berlin and Rome. When he learnt that Dino Alfieri, the Italian ambassador in Berlin at the time, was being recalled to Rome for consultations after lengthy conferences with Ribbentrop, another man 'Der Chef' detested, 'Gustav Siegfried Eins' merely guffawed. Returning for consultations to Rome? 'Der Chef' asked mockingly. That obscenity of an Italian ambassadorial womanizer has at last had his come-uppance. When one of our comrades, a fellow officer, returned from the Eastern Front not long ago – a Front incidentally, where one saw remarkably few Italians – he discovered his wife *in flagrante delicto* with the Italian ambassador. Our comrade, 'Der Chef' went on, drew his service revolver and Alfieri would by now be dead if the cringing coward of a Macaroni had not gone down on his knees and – would you believe it – pleaded diplomatic immunity. So our comrade gave the Macaroni the thrashing of his life and then bundled him off to his embassy. That was why Dino Alfieri had to return to Rome. He simply could not risk being seen in public, and his nose required plastic surgery.

The broadcast was a complete 'sib' but it was believed in Rome. Ciano noted in his diary that Alfieri's star appeared to be on the wane in Berlin, and when Mussolini, no mean womanizer himself, heard – and believed – how Alfieri had been beaten up by a German officer, he was reported to have laughed his head off.

Another 'sib' spread by 'Der Chef' which was accepted as truth by many of Germany's fighting men and quite a few members of the German public – probably because they wanted to believe it – was that top Nazi leaders like Dr. Ley, the Minister of Labour, were not subject to food rationing and that they received so-called 'Diplomat Rations'.

This foul iniquity – according to 'Der Chef' – had come to light when the father of a kitchen maid who had left Dr. Ley's employment, had telephoned Dr. Ley's butler to ask for his daughter's ration cards. The butler had told him that Dr. Ley and his household did not have ordinary ration cards because they were entitled to 'Diplomat Rations'.

The 'sib' gained such wide circulation inside Germany that Dr. Ley, unwisely, tried to deny it in no less a Nazi newspaper than *Der Angriff*. 'We National Socialists', he wrote, 'know no such thing as "Diplomat Rations". Every man, whether he is a Reich Minister or a Reich Leader, has to live on his rations just like any ordinary workman. . . . The normal rations are enough. I myself am a normal consumer and live on them.'

The effect of Dr. Ley's denial was mainly to stimulate the circulation of the 'sib' and to provoke ribald remarks about his drinking habits which were known to border on the excessive. As 'Der Chef' was quick to point out, Ley – a normal consumer: as far as bread or alcohol is concerned?

By October 1943 Delmer decided that 'Der Chef' had to be caught by the Gestapo and eliminated. His end, after more than two years, was short and dramatic. The ever-vigilant Gestapo tracked down his elusive van and interrupted 'Der Chef' in one of his broadcasts. There was a brief burst of tommy-gun fire, a snarling 'Got you, you swine', and that was it.

Unfortunately, 'Der Chef's' carefully rehearsed and recorded demise was repeated by accident. It is not known how many people heard the second broadcast; in any case it did not much matter. After all, a dead propaganda station can fortunately not lose credibility.

What was important about 'Gustav Siegfried Eins' in October 1943 at the time of its demise, was that during its operation Delmer and his team worked out the broad framework of 'black' propaganda policy which was to guide the 'Kurzwellensender Atlantik' ('Short-wave station Atlantic') and the 'Soldatensender Calais' ('Soldiers' Station Calais'), the latter backed by the most powerful medium wave transmitter in Europe.

The announcers talked in the language and idiom of the German soldier. They were patriotic, they were for Germany. But, they asked, was the same true of every German? Regrettably not. The Nazi Party bosses – the *Parteikommune* – were interested only in continuing to live off the fat of Europe as long as they could. Did they and their families stay in the towns that were being bombed nightly by the R.A.F. and the U.S.A.F.? Of course not; they were safely and comfortably installed in country houses. Did they take in bombed-out families in their country retreats? What a question! In the first place, that would have meant sharing kitchen and bathroom facilities with strangers, and secondly, and more seriously, the refugees might get a sight of the tennis courts and swimming pools on the country estates. A highly efficient intelligence organization kept 'Atlantik' and 'Calais'

19. Some of the 250,000 Axis soldiers who surrendered in Tunisia. Goebbels had called it a 'peripheral' theatre of war, but Axis troops on whom this photograph was dropped were unconvinced.

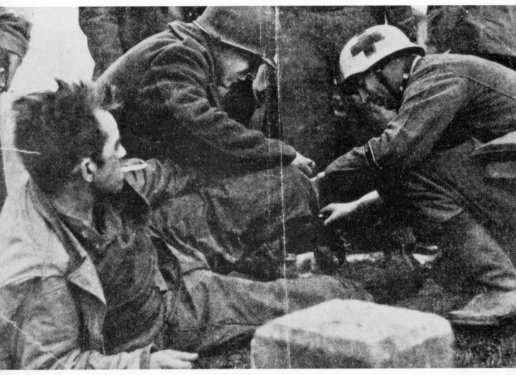

20. The caption to this picture read: 'Germans are chivalrous fighters. This severely wounded American soldier is receiving first aid before being brought in an ambulance to the hospital, where he will be treated by first-class German doctors'.

supplied with the names and addresses of the estates of Party bosses who employed special guards, accompanied by Alsatians, to keep bombed out families well away from their estates. Did the wives and daughters of the Party bosses have to work in factories? The answer was: of course they were automatically exempt.

The German front-line fighting soldier, 'Atlantik' and 'Calais' proclaimed, was brave, honest, decent and long-suffering – but what chance did he have with a corrupt Nazi Party leadership?

Delmer had, from 1941 until July 1944, refrained from allowing any of his 'black' stations from attacking Hitler directly – in the autumn of 1943 he also became responsible for 'black' propaganda to Italy, Hungary, Bulgaria and Rumania. He had insisted on the Party clique being treated as the main target to attack. After 20 July 1944, after the failure of, as he had his stations call them, the 'peace generals' assassination plot, he changed his tactics. Hitler was from then onwards a main target. The British, Americans and Russians, 'Atlantik' and 'Calais' commented, were the last people to want Germany to lose its Führer. On the contrary, they asked for nothing better than to have the Fatherland led by a man who had never learned the soldier's trade, who relied on mystic inspiration. A man like Hitler was for the Allies – an ally.

And in the months and weeks that followed, listeners to 'Atlantik' and 'Calais' were regaled with details of the Führer's mental and physical diseases, with the obscurantist cures with which a succession of witch-doctors and astrologers treated him, with broad hints that the Führer was not quite in tune with reality. No wonder Germany was losing the War. No wonder the equipment with which the ordinary German front-line soldier had to fight was not as good as it should be. In scores of stories these themes were driven home every day – and every story, no matter how slanted, was carefully built up on a foundation of accurate information supplied by intelligence or by German prisoners of war. For from 1943 onwards many Germans who fell into British or American hands were only too eager to help to bring Hitler down. A mechanical fault in a German fighter plane? A shot-down Luftwaffe pilot had supplied all information, and Luftwaffe pilots on the other side recognized it as genuine; they therefore accepted the conclusion that German aircraft production was being consistently sabotaged. Disagreement in the the German High Command? German staff officers who heard stories of this sort, often substantiated by small and accurate details, were left wondering how much had in fact leaked out. When the British Admiralty got wind that a U-boat had left Lorient, Calais played a special request record for a member of its crew 'somewhere in the Atlantic', and, as captured U-boat men explained after-wards, 'you immediately felt the eyes of every ship in the Royal Navy on you; it was like being in a submerged coffin' – very demoralizing! Yet unpleasant and unwelcome though much of its broadcast content may have been to staff officer, Party official and rank and file soldier, sailor or airman on the German side, 'Atlantik' and 'Calais' made compulsive listening. One had to know everything there was to know, and that meant listening to 'Atlantik' and 'Calais'. And even for those who did not regard listening as compulsory, 'Atlantik' and 'Calais' provided a very special listening attraction – its continuous all-night-through dance music, including the very latest records of popular music from the United States which were made available to Delmer through the good offices of the head of OSS, 'Wild Bill' Donovan. None of this music could be heard in Europe from any other station. It was a unique attraction, one the hundreds of German signallers and observers between the North Cape and the Bay of Biscay during the long night hours of duty could not resist. Nor could they resist absorbing the bits of information which were broadcast between the music, and telling their comrades about them once they came off duty.

Whatever the reason for listening, those who tuned in soon gained the impression that none of Germany's secrets seemed to be safe from the Soldatensender, and only four weeks after it first went on the air, Dr. Goebbels noted in his diary that 'the so-called Soldatensender Calais, which evidently originates in England . . . gave us something to worry about. The station does a very clever job of propaganda, and from what is put on the air, one can gather that the English know exactly what they have destroyed in Berlin!'

It was the accuracy of the details of bomb damage inflicted by Allied air-raids, as well as the essentially truthful background of much of the material, which established the credibility of 'Atlantik' and 'Calais' in the eyes even of those who were not taken in by the pretence that what they were listening to was a genuine German Armed Forces station. It was amazing, however, how many fell for the ruse, and it would be rash to dismiss them as fools. After all, what was a German listener to make of 'Calais' when an address by the Führer over all German stations had been announced, and 'Calais' broadcast Hitler as he was speaking like every other German station? What the German listener did not know was that the radio engineers working for Britain's 'black' propaganda organization had perfected a system by which 'Calais' could at will be joined to the national German radio network.

From the beginning of 1944 the 'Soldatensender' concentrated on softening the morale of the German troops who were soon to be involved in fighting the Western Allies in France. The needs of the Western Front were being subordinated to those of the Eastern Front. The troops in the West only received equipment that was second best; all the best stuff went to the East; and far too many units in the West were second-rate as well. Not their fault, of course; this was bound to happen when more and more young and old men had to be called up and sent off to France before they had completed even their basic rifle training. And just in case some of the new units should have ambitious commanders who wanted to lick them into first-class fighting material as quickly as possible, 'Calais' warned that units which showed themselves too smart and efficent would be sent to the Eastern Front. Promotion and efficiency in France, it told its listeners, was a sure way to death in Russia.

And then what about all those Ukrainian and Croat formations in France? (Such formations were in fact stationed in France to help with occupation duties.) They, the 'Soldatensender' commented sarcastically, could be relied on to run and leave the flanks of the German troops exposed the moment they heard a shot fired in anger. As for the Atlantic Wall, was it really any good? The British and the Americans had developed what could be described only as 'miracle' weapons – new armour-piercing bombs and phosphorous shells which were capable of penetrating the toughest armour and ferro-concrete and destroying every one inside. 'Calais' reported that the Americans had supplied these shells to the Russians, who had used them to eliminate the German ring of fortifications around Sebastopol; the Anglo-Americans had used these 'miracle' shells with appalling effect against German fortified positions on the Italian Front.

In tough soldiers' jargon, the 'Soldatensender' ticked off all the 'joys' to come. Air cover? There would be none. The Americans and the R.A.F. will bomb all the bridges and railways and cut off German supplies. The Luftwaffe had virtually no planes left. All it had got was the fattest and most decorated air marshal in the world. Of course Germany had one advantage; it had a surfeit of high Party bosses who considered themselves military geniuses and were constantly meddling with our officers who knew their job. By the time they stopped their meddling, many good German soldiers would unfortunately no longer be around.

The effectiveness of the 'Soldatensender' in the weeks and days before the invasion of Normandy was due largely to the way it reflected and played on the grievances and misgivings of the German armed forces in Western Europe.

The invasion of the Continent on 6 June 1944 electrified Occupied

Europe. Liberation was at hand at long last. The nightmare of Nazi occupation was about to end.

It needed no propaganda to drive home the exciting prospects opened up by the Normandy landings. The facts spoke for themselves. Far from playing up the significance of the invasion, in fact, London had still to urge restraint on the peoples of Occupied Europe. 'Don't move too soon; wait until we have broken out of the Normandy bridgehead; co-ordinate with our armies – the Germans retain the strength to retaliate viciously against you if you act too soon. Be patient; wait for instructions.' It was advice which, after years of being pent up, not all members of the Resistance movements found it easy to follow.

The restraint urged on the Resistance in Occupied Europe by the Western Allies contrasted sharply with the practice the Russians followed in their broadcast and printed appeals to underground movements. Even before the Soviet armies went over to the offensive, they urged all who could to carry out acts of sabotage no matter how great the risks. As a country fighting for survival, Russia seemed to feel justified in calling on others to bear equally heavy sacrifices – a view she continued to propagate throughout the War. In the light of Russia's enormous losses, the perhaps avoidable loss of a group of Resistance workers appeared to her a matter of comparative insignificance.

What was the effect of the Normandy landings on German morale? Since Stalingrad, Goebbels had insisted that the only way in which the Western Allies could decisively breach *Festung Europa* and end the War in their favour was by a successful assault from the West across the Channel. This was precisely what the Western Allies had done. The Germans had failed to drive them back into the sea, and although they were still containing the Allies in Normandy, it was plainly only a question of time before the Allied build-up of men and material was sufficient to enable them to break out.

This then was the moment of crisis. Was something in Germany going to give?

It was at this point, on 20 July 1944, that a group of German officers – including some who had been decorated by Hitler for conspicuous bravery – attempted, 'in the higher interests of the Fatherland', to assassinate the Führer and set up an alternative German government. The attempt failed, and with it vanished the only hope of the formation of a German government willing and capable of ending hostilities without further bloodshed.

It is sometimes urged that Roosevelt's and Churchill's declaration at

Casablanca in 1943, insisting on 'unconditional surrender' of the enemy powers, prolonged the War because it made it difficult for any anti-Hitler group to emerge in Germany, seize power, and make peace. The argument ignores the fact that Hitler had since 1933 deliberately eliminated all possible rival contenders for power. Except for the Nazi Party, all political parties of the Right and Left, and the trade unions, had been abolished and their leaders imprisoned, executed or driven into exile. There remained no person of national eminence in any walk of German life around whom men could rally in a moment of supreme crisis. Only the German army could provide a possible alternative rallying point, and with the failure of the July Plot that alternative disappeared.

The position was different in Japan. Japan was ruled by a power élite but within that élite there were factions, and first one faction and then another exercised a dominating influence. General Tojo, who was Japanese Prime Minister until the middle of 1944, was the leader of the hard-liners; Admiral Suzuki, who became Prime Minister in 1945, belonged to a more moderate faction. It was to this more moderate faction that Captain Zacharias, U.S.N., addressed himself in his broadcasts.

In Germany there was only Hitler, and after the failure of the 20th of July Plot he immediately reasserted his complete control. Every senior German commander had publicly to pledge his loyalty to him, while every soldier, whether field marshal or private, who failed to fight on to the best of his ability, was branded a traitor to his oath of loyalty and the Fatherland and executed. Himmler was made Commander-in-Chief of the German Home Army and its reserve and training units, and his men were assigned to every important command and communications post in the Wehrmacht. Like a dazed giant, the German war machine carried on because there seemed no other choice.

Rundstedt was brought out of retirement to take charge of the German Ardennes offensive in the winter of 1944, and when the offensive collapsed, he was returned into retirement; but the troops under his command had carried out their job like the disciplined, highly trained, hard-bitten soldiers they were. Their failure did not surprise them, and those that came back were re-formed on the other side of the Rhine to await the next onslaught of the Allies. The war was lost – but Hitler decreed that it had to go on. That was all there was to it.

In this situation, propaganda, German and Allied became virtually meaningless. Field Marshal Kesselring, who succeeded Rundstedt as Commander-in-Chief, West, perhaps reflected the grim mood of hopelessness, of being caught in a situation from which no escape seemed possible – a mood prevalent among many German officers at the time – when he said to his staff on arriving at his headquarters; 'Well, here I am – the V3'. Everyone present knew that Hitler's previous two 'miracle' weapons, the V1 and V2 rockets, had come too late and in too small numbers and therefore failed. Everyone, including Kesselring, also knew that the army groups which supposedly made up Germany's Command West, existed only on paper. Yet there appeared no alternative but to carry on. Those who wavered were reminded of what the Führer expected of them by the presence of Himmler's execution squads.

The only German officers who had a freedom of choice were those well removed from the watchful eyes of Himmler's minions – in harbours and fortresses along the Atlantic and Channel coasts and in surrounded towns. And in such cases Allied propaganda could, and occasionally did, perform a positive and effective role.

On 25 June 1944, Lieutenant General von Schlieben, who was in command of the key fortress dominating the approaches to Cherbourg, was asked by General M. S. Eddy, of the 9th U.S. Infantry Division, to surrender. There was no reply to this demand by General von Schlieben.

On the following day, 26 June 1944, the Americans tried a different line. Using a loudspeaker, an American spokesman addressed General von Schlieben directly on behalf of General Eddy. 'You and your men have put up a gallant fight', the spokesman said, 'but your position is hopeless. The only thing for you is to surrender while there is still time. Otherwise you and your men will be destroyed.'

The silence which followed this appeal seemed endless – although in fact it lasted but a few minutes. Suddenly the silence was broken by a German voice, speaking from the fortress in German. It was the voice of General von Schlieben. 'I cannot surrender', the General announced, 'my orders are to fight to the last man and the last cartridge. It would be different if you could prove to me that our position is hopeless. If you could, for instance, fire one of those phosphorous shells which can penetrate the heaviest armour and ferro-concrete, then I – as a responsible senior officer of the German army – would be compelled to recognize the hopelessness of my position and act accordingly.'

The Americans readily obliged. One of their batteries fired a perfectly ordinary shell at the fort. A white flag went up, the heavy steel and concrete main gates of the fortress swung back, and General von Schlieben led his men out of the fortress to surrender formally.

His surrender was a success for which the Soldatensender Calais' and the local combat psychological warfare team could both claim credit. Plainly, General von Schlieben and his men had heard the 'Soldatensender's reports about the new 'miracle' weapon, the phosphorous shells, and although as a professional soldier he must have realized that he and his men were facing no 'miracle' weapons after the perfectly ordinary shell had been fired, the local American combat psychological warfare team punctiliously respected his surrender as that of an honourable soldier facing overwhelming odds.

It was a point the Allies did not observe in every instance. At St. Malo they adopted similar tactics to those employed by the Japanese against General Wainwright at Corregidor. They tried to divide the commanding officer from his own men by implying that he would fight to the last man and then surrender to save his own skin. Now, soldiers in any army of the world have their own views about their commanding officers, but they do not like to be told by the enemy that he is a self-centred brute, even in a difficult situation. The Japanese made that mistake at Corregidor in the Philippines. The Allies made the same mistake when they called the German commander of St. Malo, Colonel von Auloch 'mad'. As a result, von Auloch held out until his food and ammunition were exhausted, then informed Hitler by radio that he and the survivors of St. Malo had done all the Führer and Fatherland could expect of them and respectfully requested that he be allowed to surrender in order to obtain medical care for his wounded. All this time the Allies were reviling the 'mad' Colonel for ruthlessly throwing away lives in a hopeless cause.

Then, suddenly, when he was good and ready, von Auloch surrendered. He introduced himself as the commander of St. Malo whom the Allies had called 'mad', but he asked his captors not to call him 'colonel' but 'general'. A grateful Führer, he explained, had promoted him because of the heroic resistance he had put up to which the Allies had given such wide publicity. The Führer, he went on, had also awarded him the 'swords' to the Knight's Cross of the Iron Cross. In short, Allied propaganda had won von Auloch almost the highest decoration for bravery in the field as well as promotion to general.

Perhaps the most tragic mistake Allied combat propaganda made in not taking into account the state of mind of the German commander who was being asked for surrender, was in the case of Aachen in October 1944. Commanders such as these may not have had Himmler's men breathing down their necks, but they all had families in Germany on whom retribution could be exacted; and the atmosphere in the Germany of 1944 and early 1945 was not unlike that prevailing among rival gangs in Chicago in the 1920's. The German commanders also considered themselves officers and gentlemen; they lived by a code of conduct which may have seemed ludicrous to outsiders, but it was a code that imposed certain standards, not least among which was regard for the welfare of one's men in the face of overwhelming odds. For anyone concerned with psychological warfare on the Allied side, to take no notice of the existence of that code, was equivalent to refusing the help of men who, by being offered honourable surrender, could have avoided further fighting in some areas and so saved Allied lives.

The Allied demand for the surrender of Aachen, addressed to its commander, General Count von Schwerin, made no attempt to salve his honour. Even so, General von Schwerin was apparently inclined to accept the terms of surrender in order to avoid senseless slaughter, but before he could do so he was relieved of his command and replaced by a well-known Nazi fanatic, Colonel Gerhard Wilck.

The bitter irony of the operation was that Wilck eventually surrendered, with not a speck of dirt on his impeccably pressed uniform or his gleaming riding boots. Aachen had, in the meantime, been reduced to rubble, and thousands of Allied and German soldiers as well as civilians had perished.

The story of the German commander of the Channel Isles, the only part of the territories belonging to the British Crown to fall under Nazi occupation, left a less bitter taste. Left behind on the Channel Isles, once the Allies had broken out of the Normandy bridgehead and pushed the Germans far to the North, he refused all demands to surrender. He calculated that the Allies knew that he and his meagre garrison could do nothing to endanger them but equally that it was not worth their while to risk the lives of their sailors and soldiers or waste the resources required for an amphibious assault on the islands he occupied. Why, therefore, surrender? It would be more comfortable to sit the War out on the Channel Isles than even in an American prisoner-of-war camp. He surrendered to the Royal Navy on 9 May 1945 – one day after hostilities with Nazi Germany came formally to an end.

All these, however, were side shows. As far as the centre piece of the drama was concerned, propaganda played a comparatively insignificant rôle from mid-1944 onwards. Given the situation which existed after the July Plot, the argument of the Allies that it was senseless to continue fighting seemed strangely irrelevant. The Germans knew that this was true, but what difference did that make? In the circumstances Germany would continue fighting either until Hitler was dead or until resistance simply disintegrated in chaos and confusion.

As it happened, the two events coincided.

CHAPTER 15

Conclusion

No battle, no campaign in the two World Wars, has been won by propaganda alone. Propaganda is no substitute for military victory. It cannot unmake defeats. It can help prepare the way for the former and speed its coming, and it can mitigate the impact of the latter, but it cannot act in isolation. To be effective, it must be closely related to the course of events.

It was highly effective, and probably decisive, in the defeat of the Austro-Hungarian armies in the Battle of the Piave in June 1918 in the First World War, but it played no part in affecting the result of the Battle of Alamein or the Battle of Stalingrad in the Second World War.

Hitler often paid tribute to Britain's and America's propaganda in the First World War. He saw it as one of the main reasons for Germany's defeat, but there is really nothing to support his view. Germany lost the First World War because her armies were defeated in the field, her industries starved of raw materials and her population of food. It no doubt suited Hitler's purpose in re-militarizing Germany in the 1930's to obscure these unpleasant facts and to make the Germans think they had been tricked into defeat by some clever propagandistic word-play on the part of President Wilson and his Fourteen Points or by Lord Northcliffe and his team at Crewe House in the heart of London's Mayfair. But propaganda by itself has never yet won a war or even a battle.

The handicap under which German propaganda suffered in both World Wars, and Japanese propaganda in the Second, was that those charged with conducting it were too concerned with the societies in which they themselves lived and with the problems and tensions of those societies to be able to make the effort to 'think themselves into the other fellow's skin'. Hence they failed in applying the four main guide lines which are essential to propaganda if it is to be successful: identifying the target; determining the message to be put across;

establishing credibility; and selecting the appropriate means of communication.

Even so astute a propagandist as Goebbels appears to have been unable to escape the myths surrounding Germany's defeat in the First World War. On 25 January 1942 he wrote in his diary; 'Our adversaries lament the fact that they have no compelling peace slogan. Quite obviously they would like to use it to deceive the German people. I will not permit this theme to be discussed by our writers because I am convinced that so delicate a problem had best be put on ice and killed by silence. We can surely congratulate ourselves that our enemies have no Wilson Fourteen Points. Of course, if they had them, we would not be duped by them as were the German people of 1917 and 1918.'

Goebbels was, of course, a product of his times, and like many Germans of his generation he may genuinely have believed that Germany was defeated in the First World War as a result of the insidious propaganda effect on the German people of President Wilson's Fourteen Points and Lord Northcliffe's sinister manipulations. Yet this same man could, with unfailing instinct and dexterity, insert the poison of disruption if the occasion arose. He did so in his handling of the Katyn incident. He accurately assessed the antagonisms and suspicions between the Russians and the Polish government in exile in London and the trouble any rupture between them would cause between the major Allies, and he exploited the situation as wickedly as he could.

At the same time, brilliant though he was as a propagandist on occasion, he may, like many Nazis, not have been very clear in his mind about the difference between propaganda and coercion. In the Nazis' struggle in winning power in Germany between the two Wars, the two were complementary elements of the same package. The object was to persuade people to support the Nazi cause. Where slogans failed, the persuading was left to the jack boots, knuckledusters and revolvers of the stormtroopers. The same method was applied in the Second World War. When nations would not accept Hitler's demands, the Wehrmacht was thrown in to make them comply, and a large part of Germany's propaganda effort was devoted to terrorizing Hitler's victims into submission by stressing the invincibility of the Wehrmacht.

In short, Nazi propaganda – even under the direction of a man as subtle as Goebbels – showed itself curiously rigid and inward-looking, dominated purely by internal German experience and, on the whole, incapable of understanding the outside world. And even within these narrow limits, it eventually proved sterile. Goebbels, in his diary, may have boasted that the German people would not be duped, as they had been in 1917 and 1918, even if the Allies had something to offer like Wilson's Fourteen Points; but in the last resort it was Hitler's execution squads that kept some semblance of organized armed resistance going until the bitter end and not Goebbels' propaganda.

In the light of these circumstances it is very difficult to support those who argue that the end of the Second World War was postponed by the demand for 'Unconditional Surrender' which Roosevelt and Churchill laid down at Casablanca in 1943. Naturally Goebbels made the most of this declaration to strengthen German resistance. He also exploited Morgenthau's alleged plan for the pastorilization of Germany and Vansittart's demands for a 'harsh peace'. But did German morale have to be stiffened in this way? Certainly not after the failure of the 20th July Plot, 1944, against Hitler, and after that, Himmler's execution squads were the determining factor.

The Allies, at least the Western Allies, were less rigid in their approach to propaganda than either the Germans or the Japanese, both of whom were prisoners of their own internal political experiences and therefore found it difficult, if not virtually impossible, to think themselves into 'the other fellow's skin'. The Soviet propaganda effort, too, was inhibited and lacked the flexibility essential to effective psychological warfare. This was due partly to the straightjacket Communist dogma imposed on Soviet propagandists and partly to the inspiration and almost atavistic antagonism Communism aroused. As a result, Moscow was not given the chance of a fair hearing even when its arguments were reasonable.

By comparison with the more disciplined countries of Germany, Japan and Soviet Russia, the propaganda effort of the two main Western Allies – Britain and the U.S.A. – was in theory chaotic but in fact probably the most effective. The organization in both countries was, to put it mildly, a muddle. Elmer Davis of the OWI was never quite certain whether either President Roosevelt or Secretary of State Hall knew what he and his agency were doing, and Sir Robert Bruce Lockhart, the head of Britain's Political Warfare Executive, commented sadly on one occasion that more of his time appeared to be taken up in fighting other British government departments than in fighting the enemy.

Moreover, these somewhat ramshackle organizations attracted a wide variety of characters from all walks of life whom even an eccentric Hollywood director would have hesitated to bring together and weld into a team. There were journalists, diplomats, actors, publishers, soldiers, bankers, exporters, painters, politicians, shop-keepers, policemen, university professors, refugees from the enemy-occupied countries, radio experts, printers, cartoonists. They all spent a good deal of time and effort fighting one another; yet all their often impassioned in-fighting resulted in their concentrating on what should be the real purpose of propaganda – on how to try to influence people how to think and, if possible, to act. The one fact that emerged clearly from these squabbles, was that people cannot be persuaded to think or behave in a certain way unless, deep down at least, some part of them wants to. Propaganda, in other words, must seek out and strike a chord that is already there.

German propaganda did not invent the political divisions and prejudices of the French body politic in 1940: it merely exploited them. The B.B.C. raised the morale of Occupied Europe in the dark days of 1940 and 1941 because, although sunk in despair and gloom, few people on the Continent wanted to live under Nazi occupation. Goebbels rallied the German public after Stalingrad because most Germans had a horror of living under Soviet rule. The success of the 'Soldatensender Calais' in shaking the confidence of Germany's West European armies and U-boat crews, was due to the fact that it skilfully played on the grievances, grouses and fears of the German serviceman.

And Captain Zacharias in his broadcasts to Japan from America would not have evoked a response if the moderates within the Japanese power élite had by the beginning of 1945 not won the upper hand over hard-liners like General Tojo and were looking for a way out of the mess in which they felt their country found itself.

Propaganda was a part of the war effort of all the belligerents in both World Wars, a valuable and at times even an essential part. But what rôle is it likely to play in future wars?

The evidence suggests that its importance will probably increase rather than diminish. Since 1945 the world has seen a series of so-called localized wars. Before, during and after the fighting the protagonists have striven or are striving hard to win the support of as much of the rest of the world for their side as they can by any and every means of propaganda.

Since 1945 there has also been no Third World War. The development of nuclear weapons may prevent such a catastrophe. In the two World Wars words were one of a variety of weapons in the armory of the belligerent powers. In future, because of the advent of nuclear weapons, words may be the only arms which the super-powers can employ without risking annihilation.

Index of Names

Index of Subjects

Entries are arranged chronologically, the space between those in sub-headings denoting the division between the First World War and the Second World War.

Aachen, surrender of German commander at, 183–4

Aeroplanes, used for dropping leaflets, 69, 79

Alamein, 157, 185

Allied propaganda, attempts to split Central Powers, 73; attempts to undermine morale of Austro-Hungarian forces, 77–80; attempts to undermine morale of German Army, 80–3;

unimpressiveness of early efforts, 107–10; defensiveness of, against German propaganda, 156; offensive against Germany, 168 ff; B.B.C. broadcasts to Germany, 168–9; 'black' radio, 170–9; and surrender of German commanding officers, 182–4

Allies, (*see also* Britain, France, Russia, Union of Soviet Socialist Republics, United States, German attempts to split, 73;

armistice with Italy, 1943, 135–6; Japanese exploitation of U.S. distrust of Britain and U.S.S.R., 148–50; German attempts to split, 161, 164–6; Poland the weak link in Soviet-Anglo-American

alliance, 163; invasion of Europe, 179–80

'America First' movement, 90

American Civil War, 66

American Revolution, 60–1

American Truth Society, 56

Anglo-American Psychological Warfare Branch, 135–6

Ardennes offensive, 181

Armenians, Turkish massacre of, 50

Asia, Japanese propaganda theme of 'Great East Asia Co-Prosperity Sphere', 151–4

Atlantic Charter, 91, 127; application of principles of, to Japan, 141–2

Atlantic Wall, 161, 166

'Atlantik' (radio station), 170, 176–9

Atom bombs, 25

Atrocities, alleged German, in Belgium, 11; Bryce Report on, 47–8, 58

Atrocity, war an, 43–4

Atrocity propaganda, 43–51, 92; inevitability of, 43–4; themes of, 44; harmful effect of defensive denials of, 46–7, 48; reactions to that of First World War, and scepticism of it in Second, 92–4

Austro-Hungarian forces, Allied attempts to undermine morale of, 77–80

Austro-Hungarian

propaganda, attempts to undermine Italian morale, 75–7

Austro-Hungary, distribution of Wilson's 'Fourteen Points' in, 23; German alliance with, 53–4; Allied attempts to split from Germany, 73; suppressed minorities in, 77–8

Balfour Declaration, 65–6

Balloons, for distributing leaflets, 83–4

B.B.C., credibility of, 17–18, 21, 113; broadcasts to Europe, 108 ff, 116, 168–70, 188; over-optimism of broadcasts to Norway, 108–9; 'Germany won't win' message of, 109; V for Victory campaign of, 110–13; messages to, from Occupied Europe, 113; effect of broadcasts of, strengthened by those of neutrals U.S. and U.S.S.R., 113–15

Belgium, German invasion of, 30–1; alleged German atrocities in, 45; their denials of, 46–7, 48; declared groundless by U.S. journalists, 57–8; Bryce report on, 47–8, 58;

German propaganda broadcasts to, 100, 102

Bezbozhnik, 121